AND
SPLICING

Frontiers in Molecular Biology

Series editors

B.D.Hames

Department of Biochemistry, University of Leeds, Leeds LS2 9JT, UK

D.M.Glover

Cancer Research Campaign, Eukaryotic Molecular Genetics
Research Group, Department of Biochemistry, Imperial College
of Science and Technology, London SW7 2AZ, UK

Other titles in the series:

Molecular Immunology

Molecular Neurobiology

Oncogenes

TRANSCRIPTION AND SPLICING

Edited by

B.D.Hames

Department of Biochemistry, University of Leeds, Leeds LS2 9JT, UK

and

D.M.Glover

Cancer Research Campaign, Eukaryotic Molecular Genetics
Research Group, Department of Biochemistry, Imperial College
of Science and Technology, London SW7 2AZ, UK

OXFORD · WASHINGTON DC

IRL Press
Eynsham
Oxford
England

British Library Cataloguing in Publication Data

Transcription and splicing—(Frontiers in molecular biology).
1. DNA replication   2. Genetic transcription
I. Hames, B.D. and Glover, D.M.   II. Series
574.87'3282   QH450

ISBN 1 85221 080 X Hbk
ISBN 1 85221 076 1 Pbk

Typeset by Infotype and printed by
Information Printing Ltd, Oxford, England.

Preface

Our knowledge of the mechanisms of eukaryotic RNA synthesis and processing and their regulation is increasing at a rapid rate. This book brings
together leading researchers to review recent developments and to present a comprehensive account of our current understanding. The first
chapter covers the initiation of transcription, concentrating upon RNA
polymerase II and its interaction with eukaryotic transcription factors.
The regulation of pol II transcription by *cis*-acting regulatory elements,
notably enhancers, and the possible mechanisms of action of enhancers
in regulating tissue- and/or host-specific gene expression is the topic of
Chapter 2. The third chapter describes transcription termination for pol I,
II and III genes, the 3' end processing of mRNA and the role of these
phenomena in gene regulation. Finally, Chapter 4 concentrates on post-
transcriptional processing, providing an up-to-date account of RNA splicing including the role of small nuclear ribonucleoprotein particles.

We hope that the book will serve not only as a source of information
but only as a stimulus for the cross-fertilization of ideas within these
related areas. We would like to thank the authors for their hard work
in writing their contributions and especially for their patience and understanding during editing and publication.

David Hames
David Glover

Contributors

Peter Gruss
Max-Planck-Institute of Biophysical Chemistry, Department of
Molecular Cell Biology, PO Box 2841, Göttingen D-3400, FRG

Antonis K.Hatzopoulos
Max-Planck-Institute of Biophysical Chemistry, Department of
Molecular Cell Biology, PO Box 2841, Göttingen D-3400, FRG

Adrian R.Krainer
Cold Spring Harbor Laboratory, Box 100, Cold Spring Harbor,
NY 11724, USA

Nicholas B.La Thangue
Laboratory of Eukaryotic Molecular Genetics, National Institute for
Medical Research, The Ridgeway, Mill Hill, London NW7 1AA, UK

Tom Maniatis
Department of Biochemistry and Molecular Biology, Harvard
University, Cambridge, MA 02138, USA

N.J.Proudfoot
Sir William Dunn School of Pathology, University of Oxford, South
Parks Road, Oxford OX1 3RE, UK

Peter W.J.Rigby
Laboratory of Eukaryotic Molecular Genetics, National Institute for
Medical Research, The Ridgeway, Mill Hill, London NW7 1AA, UK

Uwe Schlokat
Centre for Molecular Biology, University of Heidelberg,
Im Neuenheimer Feld 282, 6900 Heidelberg, FRG

E.Whitelaw
Sir William Dunn School of Pathology, University of Oxford, South
Parks Road, Oxford OX1 3RE, UK

Contents

Abbreviations

AFP	α-fetoprotein
AIDS	acquired immune deficiency syndrome
ALV	avian leukosis retrovirus
AP	activator protein
ASV17	avian sarcoma virus 17
BLE	basal level element
BPV	bovine papilloma virus
CBP	CAT binding protein
CDP	CCAAT displacement protein
CMV	cytomegalovirus
CRE	cAMP response element
CREB	CRE binding protein
CTF	CCAAT box transcription factor
DCR	dominant control region
EC	embryonal carcinoma
EGF	epidermal growth factor
GRE	glucocorticoid response element
HBV	hepatitis B virus
HSE	heat shock regulatory element
HSTF	heat shock transcription factor
HSV	herpes simplex virus
IE	immediate early
IFN	interferon
IGS	internal guide sequence
IRE	interferon regulatory element
LPS	lipopolysaccharide
LPV	lymphotropic papovavirus
LTR	long terminal repeat
MHC	major histocompatibility complex
MLP	major late promoter
MLTF	major late transcription factor
MoMLV	Moloney murine leukemia virus
MRE	metal response element
MSV	murine sarcoma virus
MTIIA	metallothionein IIA

MTV	mouse mammary tumor virus
NF	nuclear factor
OTF	octamer binding factor
Pol I, II, III	RNA polymerase I, II, III gene
pTP	precursor terminal protein
Pu	purine
SAR	scaffold attachment region
SV40	Simian virus 40
TF	transcription factor
tk	thymidine kinase
TPA	12-*O*-tetradecanoylphorbol-13-acetate
UAS	upstream activation site
URR	upstream regulatory region
USF	upstream stimulatory factor

Trans-acting protein factors and the regulation of eukaryotic transcription
Nicholas B.La Thangue and Peter W.J.Rigby

1. Introduction

Transcription is a major level at which gene expression is regulated in eukaryotes. Elucidation of the mechanisms that control the initiation of transcription is central to an understanding of cellular differentiation and the development of the organism as a whole. In this chapter we review current knowledge of the proteins involved in transcriptional control in eukaryotes. We concentrate our discussion on those regulating the synthesis of mRNA by RNA polymerase II, the enzyme responsible for transcribing protein-coding genes, because progress in this area is presently very rapid. We will briefly discuss control processes in RNA polymerase I and III transcription and refer the reader to more thorough reviews for detailed consideration of these areas (1,2). We emphasize that such a diverse and progressive subject necessitates that we select representative examples that illustrate general principles, rather than attempt to discuss the whole field. We have also elected not to discuss the regulation of hormone-responsive genes, since this topic has been recently reviewed (3).

2. Transcriptional activation is a two-step process in eukaryotes

Transcriptional activation of eukaryotic genes is a two-step process because most genes exist at some time in an inert state tightly packaged with histones into chromatin. Before transcription can occur this inert structure must be decondensed so that transcriptional control sequences are made available to regulatory proteins. The mechanisms responsible for decondensing chromatin are unclear but could involve binding to the

nucleoskeleton through DNA sequences that function as attachment sites
(4,5), so that the appropriate topology can be adopted and the relevant
sequences exposed. Once decondensation has occurred, the scene is set
for the second activation step: the interaction of regulatory proteins,
commonly referred to as transcription factors or *trans*-acting factors, with
specific DNA sequences. Such interactions mediate transcriptional
activation. It is the identification and characterization of this latter group
of proteins that has recently progressed such that we are now able to
associate a particular protein species with binding to a given regulatory
DNA sequence. We will discuss these regulatory proteins in depth, dealing
first with the polymerases themselves and with general transcription
factors, that is, factors that recognize minimal promoter sequences utilized
by a wide variety of promoters. We then progress to a discussion of
promoter-specific factors and finally to enhancer-binding factors, using
the Simian virus 40 (SV40) enhancer as a paradigm.

3. Structure of eukaryotic RNA polymerases

All three RNA polymerases have been purified to apparent homogeneity
from a variety of sources. When analyzed by SDS – polyacrylamide gel
electrophoresis such preparations are seen to contain multiple polypeptides
but, in the absence of a dissociation/reconstitution system, one cannot
be sure that each polypeptide is required for enzymatic function. In several
cases polypeptides of the same apparent molecular weight are found in
two or all three of the polymerases, raising the possibility that the enzymes
share common subunits. In one case this possibility has been shown
to be true. Cloning of the gene encoding one of the subunits of
Saccharomyces cerevisiae polymerase I and subsequent gene disruption
experiments have shown that this subunit is required for the assembly
of functional polymerases I and III but not II (6). [RNA polymerases I,
II and III are sometimes referred to as A, B and C respectively.] Because
of the powerful genetic systems available, similar analyses of other yeast
RNA polymerase genes are likely to be highly informative. In view of
recent data, discussed below, which demonstrate that particular
transcription factors can function in both polymerase II and polymerase
III transcription, it would not be surprising if there were other subunits
common to more than one polymerase.
   The genes encoding the largest subunits of all three polymerases have
been cloned. In each case the N-terminal portion of the encoded protein
has extensive homology to the β' subunit of *Escherichia coli* RNA
polymerase (7 – 9). Moreover, the gene encoding the second largest
subunit of polymerase II is homologous to that encoding the β subunit
of *E.coli* polymerase (10).
   The gene encoding the largest subunit of polymerase II, but not those

encoding the largest subunits of polymerases I and III, encodes a C-terminal extension (7,9) not found in the prokaryotic polymerases which comprises a highly evolutionarily conserved tandemly repeated hepta-peptide, Tyr-Ser-Pro-Thr-Ser-Pro-Ser. Manipulation of the yeast and mouse genes shows that a minimum number of these repeats must be present in the gene for cell viability. Moreover, the repeats in the yeast gene can be replaced by some of the analogous sequences from other organisms, suggesting that they interact with similarly evolutionarily conserved structures (11 – 13). It has been speculated (11) that the encoded heptapeptide, which has considerable hydrogen bonding potential, might be involved in interacting with the highly acidic activation domains found in yeast transcription factors (see below).

4. General transcription factors

RNA polymerase II alone cannot accurately initiate transcription *in vitro* on purified templates unless supplemented with a number of general transcription factors that provide recognition of and specificity for minimal promoter sequences (14). In the promoters that have been most intensively studied, the sequences required consist of the TATA motif and associated cap site but exclude upstream sequences (5′ to the TATA) necessary for maximal transcription, such as the Sp1 binding sites and CCAAT boxes discussed below. That the TATA motif is shared by many promoters emphasizes the generality of these transcription factors, but it should be noted that a large number of promoters, including those of many 'housekeeping' genes, lack TATA boxes. This latter class of promoters is very GC-rich, thus often containing Sp1 binding sites, and its members are often found in methylation-free islands (15).

4.1. TFIIA, B, D and E

The minimal promoter sequences of the human adenovirus major late promoter (MLP), which extend from nucleotides − 56 to + 33 but exclude the major late transcription factor/upstream stimulatory factor (MLTF/USF) binding site, were used to define at least five chromatographic fractions (TFIIA, B, C, D and E in *Figure 1*) from HeLa cell extracts which, by reconstitution, are required for formation of a pre-initiation complex (14,16). One of these, TFIIC, was identified as poly(ADP-ribose) polymerase and was found to be unnecessary as the purity of the other components was improved (16). For the most part, the function of the other factors remains to be determined but some clues are available as to their mode of action. This is particularly true for TFIID, which is the only general transcription factor that binds specifically to DNA, recognizing the TATA box and surrounding sequences which can include the initiation site. Other upstream promoter binding proteins could interact

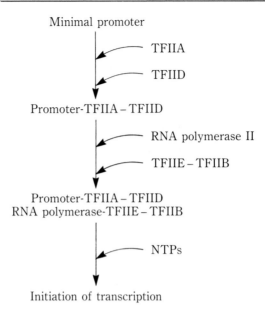

Figure 1. Proposed assembly order of general transcription factors onto minimal promoter sequences.

with TFIID and this appears to be the case for MLTF/USF, the binding of which is enhanced by the presence of TFIID, possibly because of protein – protein contacts (17,18). In addition, TFIID binding prevents nucleosome-mediated repression of promoter activity *in vitro* by an unknown mechanism which may involve changes in chromatin stucture over the entire length of the promoter (19). Such alterations, in concert with protein – protein interactions involving TFIID, could then focus factors onto upstream promoter motifs.

The assembly of TFIIA, B, D and E onto minimal promoter sequences forms a heparin-resistant pre-initiation complex that can initiate transcription upon addition of nucleoside triphosphates (20 – 22). Kinetic studies, combined with order-of-addition experiments, suggest that TFIIA acts at an early stage in the formation of the pre-initiation complex followed by sequential interaction with TFIID and RNA polymerase II [in the absence of TFIIE and TFIIB (20 – 24)], creating a complex resistant to low levels of the detergent Sarkosyl®. The formation of a pre-initiation complex stable to heparin requires addition of TFIIE and TFIIB (24).

Although TFIIA acts first in the initiation process, it has, in contrast to TFIID, a non-specific rather than a specific binding activity (24). TFIIA may therefore operate in a manner such that binding to DNA induces structural changes which allow TFIID to bind. TFIIA was originally thought to be related to actin (25), but when purified consists of three polypeptides of 12 – 20 kd (26).

TFIIB has been purified from HeLa cells and the activity shown to reside in a 30 kd polypeptide (27). By sedimentation analysis, TFIIB binds to TFIIE, and possibly weakly to RNA polymerase II (27), in contrast to TFIIE, which clearly binds to polymerase II (27). The TFIIB activity characterized in these studies is probably related to the 27 kd RNA polymerase B transcription factor 3 (BTF3) (28), although BTF3 binds to polymerase. Both TFIIB and BTF3 are active on a wide spectrum of promoters (27,28). In addition, proteins called Rap30 and Rap72, purified by affinity chromatography on columns containing immobilized calf thymus RNA polymerase II, may also correspond to TFIIB and/or TFIIE (29).

A model describing the sequentially ordered assembly of these general transcription factors is shown in *Figure 1*. Initially, TFIIA influences the DNA in some unknown way to facilitate the binding of TFIID into a complex resistant to low concentrations of Sarkosyl®, allowing promoter recognition by RNA polymerase II. This is followed by TFIIE and TFIIB binding to the polymerase, generating a pre-initiation complex capable of responding rapidly to the addition of nucleoside triphosphates.

4.2. Other general transcription factors

Several other associated factors have been identified, although their generality remains to be determined. One of these, a 38 kd protein called SII, isolated from Ehrlich ascites cells, stimulates RNA polymerase II in a random transcription assay, possibly by binding to it. Antibodies specific for SII block transcription in isolated nuclei (30,31). This factor is most probably equivalent to TFIIS, a factor that stimulates the rate of elongation by increasing the efficiency with which the RNA polymerase II passes through pausing sites (32). SII can exist both in phosphorylated and dephosphorylated states (30) but does not appear to be an essential component of the RNA polymerase transcription machinery *in vitro* (31). The activity of another factor, IIX, defined by sequences downstream from the MLP cap site, stimulates *in vitro* transcription (24), but again the mechanism of action is unknown.

Evidence for general transcription factors affecting upstream promoter binding proteins has come from studies on the chicken ovalbumin promoter for which two factors have been characterized, an upstream promoter factor called COUP and a general transcription factor S-300II (33). The purified COUP factor of 45 – 50 kd binds to a defined sequence motif (–70 to –85), and this binding is enhanced by the purified 39 kd S-300II factor in a manner that does not involve the latter binding to DNA (34). Because S-300II also had this effect on promoters that do not bind COUP, such as those of the mouse mammary tumor virus and the lysozyme gene, it may well fall into the category of general transcription factors that act outside minimal promoter sequences. It most probably mediates the effect through binding to sequence-specific promoter binding proteins, perhaps by prolonging the half-life of DNA – protein complexes as has been shown for COUP (34), enabling the S-300II – sequence-specific promoter binding protein complex to interact with RNA polymerase II and increase the rate of initiation. It is presently unclear how general is the action of S-300II, but it would not be surprising if proteins with a similar mode of action were specific for different families of sequence-specific binding proteins.

   We have here summarized current knowledge of the properties of general transcription factors and where they may function in the pre-initiation complex. It should be clear that at present this information is limited, although with the purification and characterization of some of these factors rapid progress is likely.

5. Promoter-specific transcription factors

Eukaryotic promoters are composed of many *cis*-acting control motifs. Some of these, such as the TATA box, are shared by many promoters and, as we have discussed, bind general transcription factors, whereas others, such as Sp1 binding sites and the CCAAT box, are found in selected promoters. The main characteristics of promoter motifs are distance-dependent activation and the requirement for location 5′ to the cap site. We emphasize that there is no clear division between enhancer and promoter motifs because certain motifs with promoter activity in one situation can be important functional enhancer elements in another, an example being the octamer sequence discussed below (Section 5.2.1). Additionally, tandem copies of some promoter motifs can acquire properties of enhancers, arguing that it is the context of a given sequence motif that determines its functional characteristics rather than them being an intrinsic property of the binding protein.

5.1. Sp1

5.1.1 Sp1 activates the SV40 early promoter
The SV40 early promoter consists of two overlapping promoters which control initiation at the early – early and late – early initiation sites. Promoter mapping studies defined the DNA sequences involved as consisting of a TATA box, which ensures accurate initiation at the early – early start site, and six tandemly arranged GC boxes (I to VI) in the degenerate 21 bp repeats. Each GC box contains the sequence motif GGGCGG, subsequently shown to bind the promoter-specific transcription factor Sp1 (35,36), although in the SV40 early promoter, GC box IV does not bind Sp1 *in vitro*, possibly because of steric constraints (35,37,38). Sp1 was originally identified through its ability to activate initiation on addition to a HeLa whole cell extract (36); this activation is selective

because addition of Sp1 to a transcription reaction containing both the SV40 early promoter and adenovirus MLP activates SV40 but inhibits MLP transcription by 40% (36). There appear to be 5000–10 000 Sp1 molecules per HeLa cell, and purification by oligonucleotide affinity chromatography shows Sp1 to consist of two polypeptides of 105 and 95 kd, both of which have the ability to bind to the GC box (39,40).

5.1.2 Sp1 binding sites in other promoters

The Sp1 binding site occurs in many other viral and cellular promoters where it may act together with other promoter motifs such as the CCAAT box. For example, the herpes simplex virus (HSV) thymidine kinase (tk) promoter has two Sp1 binding sites surrounding a CCAAT box and all three elements are required for maximal promoter activity (41). The mouse dihydrofolate reductase promoter and the human metallothionein IIA promoter have three and one binding sites respectively (42,43). Sp1 binding sites usually occur in either orientation and promoters that contain appropriately positioned sites are usually stimulated 10- to 50-fold (43,44). It is usually the site closest to the gene that mediates the strongest transcriptional stimulation *in vitro*. For example, the HSV immediate early (IE) 3 promoter has five Sp1 binding sites (from −70 to −255), but a truncated IE 3 promoter that retains one site is still stimulated 20-fold by the addition of Sp1 (45).

5.1.3 Cloning of the Sp1 transcription factor gene

A cDNA encoding part of Sp1 has been cloned by determining the amino acid sequence of Sp1 peptides and screening a HeLa cell cDNA library with synthetic oligonucleotides encoding these peptide sequences (46). The sequence of 696 amino acids from the C-terminus shows it to contain a DNA binding domain containing three Zn(II) fingers, a feature common to other transcription factors such as TFIIIA (see Section 12), and is consistent with the fact that Sp1 requires Zn(II) to bind to DNA.

5.2 CCAAT box transcription factors (CTF)

As with the Sp1 binding site, the importance of the frequently occurring CCAAT box was assessed by promoter mapping studies. This motif is usually located in a similar position to Sp1 sites, about 40–100 bp upstream, and is required for efficient transcription. The CCAAT box can occur either alone or in the presence of other motifs, for example Sp1 binding sites in the tk promoter (41) and heat shock regulatory elements (HSE) in the heat shock protein 70 gene promoter (47,48). The CCAAT box is functional in both orientations, and factors that bind to this sequence can co-operate with those binding to surrounding motifs in the tk promoter and the *Xenopus* hsp70 promoter.

It has become abundantly clear that a family of transcription factors recognize the CCAAT motif, and here we discuss current knowledge regarding these. It seems probable that the existence of families of transcription factors recognizing a single motif is not unique to the CCAAT sequence but, rather, will be a general feature of transcriptional control in eukaryotes.

5.2.1 The relationship of CTF to nuclear factor I (NFI), a protein required for adenovirus DNA replication

The CTF was originally defined by studying the transcription of the HSV tk promoter in infected and uninfected cell extracts (41). Analysis of CTF binding sites in a variety of promoters indicated that the sequence matched a portion of the recognition sequence for NFI, a protein required for the initiation of adenovirus DNA replication *in vitro* (49,50).

The initiation of adenovirus DNA replication involves covalent attachment of the viral precursor terminal protein (pTP) to the first deoxynucleotide residue (a cytidine 5′ monophosphate) of the new DNA strand. In addition to pTP, initiation of replication requires the adenovirus DNA polymerase, viral origin sequences and a number of cellular proteins that bind to origin sequences, including NFI (51). NFI stimulates the levels of the pTP – dCMP complex and there is evidence that in order to do this it must cooperate with another cellular protein, NFIII, the recognition sequence for which, ATGCAAAT, is located close to that of NFI (51,52). This sequence motif, the octamer, is also required for efficient transcription by a number of cellular promoters (see below). The NFI/CTF binding site is TGGCT(N$_3$)AGCCAA, and NFI/CTF, purified using either the high affinity origin site or a canonical CCAAT box, consists of a family of polypeptides of 52 – 66 kd able to stimulate both *in vitro* transcription and DNA replication. It seems in the case of NFI/CTF that the related proteins are able to participate in both DNA replication and transcription, as has previously been shown for the SV40 large T-antigen.

5.2.2 A family of transcription factors recognize the CCAAT motif

Another factor, called CAT binding protein (CBP), binds to the CCAAT box in a variety of promoters. Rat liver CBP clearly differs from CTF in its biochemical properties, such as heat stability, and it does not generate an identical footprint (53,54). For example, the single nucleotide change CCAAT to GCAAT enhances CBP binding but diminishes CTF binding (53). This CBP binding factor, also called EBP 20 (54), has been purified to homogeneity and is a 20 kd polypeptide which not only binds to the CCAAT motif but also to the core sequence present in many enhancers (54).

At least two other factors bind to the CCAAT motif. One, called NF-Y, binds to the opposite orientation ATTGG motif in the Y box of major histocompatibility complex (MHC) class II genes (55,56). The Y box, positioned at about − 50, together with the X box, is an important promoter domain of all class II genes (55,56). By a number of criteria, NF-Y is distinct from CTF and CBP, and forms a complex of 200 kd or larger (57). These same workers provided evidence for a fourth binding factor, NF-Y*, and other so far uncharacterized CCAAT box binding activities have been described (56,58). It is clear from these and more recent studies with purified proteins (199,200) that a multiplicity of factors recognize the CCAAT motif.

5.2.3 CCAAT box binding factors can be influenced by development

Transcription of sea urchin sperm histone H2B genes is developmentally regulated; they are expressed exclusively during spermatogenesis and not, for example, in the blastula or gastrula (59). The sperm H2B promoter has two important functional motifs, the CCAAT box and the H2B consensus octamer (59). Although extracts from different developmental stages and tissues have a CCAAT box binding factor, the binding is prevented in tissues that do not express the gene. This occurs via a displacement factor, referred to as CCAAT displacement protein (CDP; ref. 60), that binds with high affinity to sequences overlapping the CCAAT motif. This type of displacement factor, in combination with the multiplicity of CCAAT binding factors, suggests that common motifs such as the CCAAT box could be important regulatory elements in developmental and tissue-specific transcription.

5.3. Octamer binding factors

The octamer motif ATGCAAAT is a *cis* control sequence that occurs in a variety of promoters. Although originally defined in promoters (61), it is now clear that this element can be an important motif in enhancers, for example that of SV40 (62). Multimers of this sequence possess enhancer activity and cell specificity (63), indicating that the octamer motif can be either an enhancer or upstream promoter motif. In histone H2B promoters this sequence is located in reverse orientation about 40 – 50 nt upstream from the TATA box, where it mediates the induction of transcription from the G1 to the S phase (61,64). The octamer sequence occurs in opposing orientation in the heavy and light chain immunoglobulin promoters and confers B-cell-specific expression (see the following section and Section 7.1). It is also found in the enhancers of small nuclear RNA genes (65,66). The octamer motif is also important for transcriptional activation by viral *trans*-activators, for example, the induction of HSV gene transcription by the virion *trans*-activator Vmw 65 involves the formation of a complex between the octamer binding factor and Vmw 65 (67,68).

5.3.1 Octamer binding factors can be ubiquitous or cell specific

The cell-cycle-specific transcription of human histone H4 and H2B genes can be reproduced *in vitro* in extracts prepared from synchronized HeLa cells (69,70), in which they are transcribed efficiently in S phase extracts but poorly in G1 extracts (70). The octamer motif is required for the transcriptional induction in S phase extracts, whereas disruption of this motif has no effect on basal level transcription in G1 extracts (64,71). A ubiquitous octamer binding factor of 90 kd (OTF1), purified from HeLa cell extracts by oligonucleotide affinity chromatography, binds to this motif and stimulates transcription from the H2B promoter when added to G2 HeLa cell extracts (71). It is unclear how OTF1 mediates cell-cycle-specific induction, but this may involve regulation of OTF1 synthesis or post-transcriptional modification.

The same octamer sequence is present in the inverted terminal repeat of a number of adenovirus genomes, for example Ad2 (51,52,72). The protein binding to this sequence, NFIII, together with NFI, stimulates adenovirus DNA replication *in vitro* by potentiating the formation of the DNA replication complex (52). NFIII has been purified by oligonucleotide affinity chromatography and shown to be 95 kd, and is therefore likely to be closely related, or identical, to OTF1. Thus, as in the case of NFI/CTF, the ubiquitous octamer binding factor is capable of stimulating both transcriptional initiation and DNA replication.

As mentioned earlier, the octamer sequence is present in both the immunoglobulin light and heavy chain promoters and has convincingly been shown to confer lymphoid-specific expression *in vivo* (73,74), suggesting the existence of a lymphoid-specific octamer binding protein. Extracts derived from B cells contain cell-specific octamer binding proteins and transcribe the light chain promoter *in vitro* in a similar fashion to that *in vivo* (75–77). This assay enabled purification of a lymphoid-specific octamer binding factor (OTF2), shown to consist of three polypeptides, of 62, 61 and 58.5 kd (78). OTF2 activates transcription of the light chain promoter in non-lymphoid cell extracts by binding to the octamer motif where its footprint is indistinguishable from that caused by OTF1 binding (71,78).

Additional octamer binding factors have been defined which are cell-specific (Oct B1A, Oct B1B and Oct B2) or ubiquitous (NFA1, OBP100) (79,80), although it is at present unclear if these affect transcriptional initiation.

One of the key questions regarding the family of octamer binding proteins is what features of a promoter determine whether ubiquitous or cell-specific factors bind. It may be that the position of the octamer motif is the critical feature since in the H2B promoter it is located between -50 and -40, in contrast to -66 to -59 in the immunoglobulin light chain promoter. This location may determine how other proteins, for example TFIID, interact with octamer-binding proteins. If only the ubiquitous factor can cooperate with surrounding proteins, and if cooperation is required for efficient DNA binding, then the distance between motifs would be paramount in determining which octamer factor bound.

Interestingly, the octamer motif is not restricted to RNA polymerase

II promoters but appears to be an important functional *cis* motif in a number of RNA polymerase III promoters, for example those of the U6 small nuclear RNA gene and the 7SK RNA gene (see below).

5.4 MLTF/USF

The major late promoter of adenovirus has been extensively studied because it is a strong promoter *in vitro* (81). By mutational analysis two control motifs were defined, comprising the TATA box and an upstream sequence centered around -66 to -51 (82). A HeLa cell factor, termed major late transcription factor (MLTF) or upstream stimulatory factor (USF; 17,83), binds to the upstream site, the consensus motif being GGCCACGTGACC. This motif functions in a bidirectional manner both *in vitro* and *in vivo*. MLTF/USF may act synergistically with TFIID because binding of MLTF/USF enhances the binding of TFIID (17).

A single polypeptide of 46 kd molecular weight possesses the binding and transcription activation properties of MLTF/USF (84). The binding motif occurs in a number of cellular promoters, for example, that of mouse and rat γ-fibrinogen (at -80; ref. 85) and metallothionein 1 (at -95; ref. 86). In both promoters mutation of this sequence affects the magnitude of binding rather than inducibility (85,86). This is clearly illustrated by rat α-fibrinogen, which lacks an MLTF/USF binding site; transcription of the α- and γ-fibrinogen genes is coordinately controlled but the magnitude of the response differs. Thus, the MLTF/USF motif binds a transcription factor required for basal level expression in cellular promoters.

5.5 Heat shock transcription factor (HSTF)

All major heat shock genes contain multiple HSE with the consensus sequence C$-$$-GAA-$$-TCC-$$-$G. The HSE upstream promoter motif was defined by deletion analysis and will confer heat inducibility on a heterologous promoter (87). In some situations this motif can, together with other promoter motifs, act as an inducible enhancer (88). The HSE binds the HSTF, and purified HSTF alone binds to the motif *in vitro* (89,90). Initially yeast and *Drosophila* HSTFs were reported to be 70 kd polypeptides (89) but subsequent reports show HSTF to be 150 kd; the 70 kd polypeptide is produced by mild proteolysis (90). That HSTF is involved in the heat shock response *in vivo* is supported by the fact that its affinity for variant HSE *in vitro* correlates with the ability of these elements to support heat inducibility *in vivo* (90).

Inhibitors of protein synthesis do not prevent transcriptional induction of the heat shock response, suggesting that HSTF is activated by post-transcriptional mechanisms, and although HSTF is involved in the response in HeLa cells and yeast, the mechanisms are different. In HeLa cells heat shock causes more HSTF to bind to the HSE, whereas in yeast

HSTF binds constitutively and heat shock causes HSTF to become phosphorylated (91). In the absence of heat shock the TATA box of *Drosophila* heat shock gene promoters is continually occupied by a factor (92) that presumably interacts with HSTF during heat shock, thus favoring a rapid response. HSTF bends DNA on binding (93), possibly bringing the constitutively bound factor into close proximity with other proteins and thus enabling the rapid formation of an active transcription complex.

5.6 Cyclic AMP-response element binding factor

Cyclic AMP (cAMP) is a major second messenger in eukaryotic cells, where it mediates its effects by activating cAMP-dependent protein kinases that phosphorylate key cytoplasmic and nuclear molecules. Promoter mapping studies have identified in cAMP-responsive genes a DNA sequence, called the cAMP response element (CRE), that confers cAMP inducibility on heterologous promoters (94). In the pheochromocytoma cell line PC12 the protein binding to CRE has a molecular weight of 43 kd and is phosphorylated *in vitro* by the catalytic subunit of cAMP-dependent protein kinase (95). Treating PC12 cells with forskolin, an inducer of cAMP synthesis, also results in increased phosphorylation and this correlates with transcriptional induction of cAMP-dependent genes (95), suggesting that phosphorylation of this protein is necessary for it to activate transcription. The CRE binding protein (CREB) binds to the sequence TGACGTCA present in other promoters, for example, c-*fos* (−60) and several adenovirus early promoters (96).

6. Enhancer binding factors

The characteristics of enhancers as revealed by the analysis of transcription *in vivo* are considered in detail in Chapter 2. We will discuss current knowledge of the factors that bind to enhancer motifs. As more enhancers have been analyzed it has become abundantly clear that there is no clear distinction between enhancer and upstream promoter motifs, but rather that there is an 'enhancer effect' which conforms to a number of strict criteria (see next chapter). This effect can be exerted either by combinations of various motifs, as in the SV40 enhancer (62,97), or by artificially creating tandem repeats of individual motifs, for example, the octamer sequence (63,98 – 100). The octamer is an example of a motif that can be part of an enhancer or be an integral upstream promoter motif, as in the histone H2B promoter (61,62). For the purpose of this discussion, we will refer to DNA sequences shown *in vivo* to have an enhancer effect as 'enhancers'.

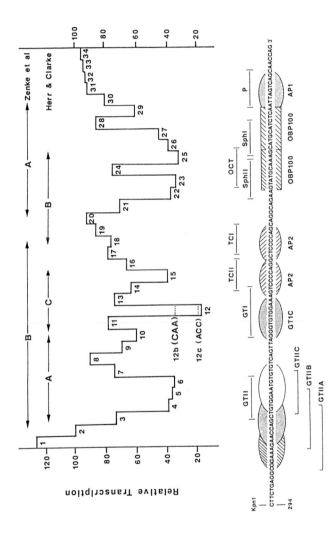

Figure 2. Protein binding sites in the SV40 enhancer. The sequence of the enhancer region of wild-type SV40 containing the distal 72 bp element and its 5' flanking region is shown. Also shown are the locations of sequence motifs identified in the enhancer region organized in two domains, A and B, by Zenke et al. (97) or three, A, B and C, by Herr and Clarke (99). The protein binding sites of nuclear factors from HeLa cells are indicated and are discussed in detail in the text. The graph shows the effect of mutations within the enhancer region on transcription from the EES of the SV40 promoter, taken from Zenke et al. (97). The relative transcriptional activity of mutants pA1 to pA34 (relative to the wild-type) following transient expression in HeLa cells is indicated. Two additional motifs discussed in the text, the KID-box and Pu-box, located within the region 346 to 294 (103), are not indicated on this diagram.

6.1 Enhancers are modular

The SV40 early control region has been subject to intense study and is perhaps the best characterized eukaryotic transcriptional control sequence. The combinatorial and modular nature of its organization is typical of many other transcriptional control elements. The early control region consists of two sets of sequences that activate transcription in *cis*, known as the early promoter and enhancer. The early promoter binds the transcription factor Sp1 and weak binding to this region is also observed for other factors such as AP-1, AP-2 and AP-4 (see Section 6.3). The enhancer, classically defined by its ability to activate transcription of heterologous promoters independent of distance and orientation (101), includes the 72 bp duplicated sequence, which accounts for most enhancer activity, and additional sequences located on the late side of these repeats (97) (*Figure 2*). Modules or functional domains of the SV40 enhancer were originally defined by either systematic mutagenesis (97,102) or point mutation revertants (98–100). Point mutation revertants detected three separate modules, A, B and C (99), that function autonomously when present as tandem copies and additionally show cell-specific activity (63). Systematic mutational analysis divided the enhancer initially into two distinct domains, A and B, that had little activity alone but which in combination resulted in 400-fold induction, irrespective of their orientation and, to some extent, of the distance between them (97). Enhancer activity could also be generated by duplication of either domain, but to a level lower than that of the wild-type. Domains A and B were further subdivided into GTII, GTI, TCII, TCI, SphII, SphI and P motifs (97). These data are summarized in *Figure 2*. [Other domains outside the 72 bp repeats, such as the Pu box and KID box, have also been defined (103) but are not shown in *Figure 2*.] These sequence motifs occur variably in other enhancers, and to some extent in promoters, suggesting that enhancers are mosaics of a limited number of basic sequence motifs and that it is the combination of these motifs that determines the cell type specific activity of a particular enhancer. In the next section we discuss our current knowledge of the proteins that recognize these sequences.

6.2 Enhancer activity *in vitro*

The enhancer effect is most marked *in vivo* but some stimulation by enhancer sequences is also observed *in vitro*. In HeLa cell extracts the SV40 enhancer stimulates transcription, usually about 10-fold, when positioned relatively close to the initiation site (104,105). In the 3′ location, however, its effect can actually be detrimental to promoter activity (106). *In vitro* transcription assays performed with competing enhancers showed, originally for the SV40 enhancer but subsequently for others, that *trans*-acting factors were necessary for enhancer activity (107,108).

There has been variable success in establishing *in vitro* activity with other enhancers and in all cases to date the stimulation *in vitro* is much lower than that obtained *in vivo*. The immunoglobulin heavy chain enhancer preferentially stimulates transcription in B cell extracts (108,109), the most marked effect so far being 15-fold (108). Interestingly, regulatory sequences that reduce the heavy chain enhancer stimulation *in vivo* also function *in vitro* (110 – 113). It is unclear why enhancer effects cannot be quantitatively reproduced *in vitro* but numerous possibilities can be envisaged, for example, a requirement to attach to some higher order nuclear structure not retained in cell extracts.

6.3 Enhancer binding proteins and the SV40 enhancer

The SV40 enhancer can be divided into a number of motifs the mutation of which reduces wild-type enhancer activity. These motifs are shown in *Figure 2*. We now discuss the proteins that bind these sequence motifs.

6.3.1 P motif

The P motif binds the transcription factor AP-1 (114) (*Figure 2*) originally defined through its selective binding to *cis* control sequences in the SV40 and metallothionein IIA (MTIIA) enhancers (114). The consensus AP-1 binding site is $TGANT^C/_AA$ and the mutations that alter this motif in the SV40 enhancer only have modest effects in HeLa cells (97), although a more pronounced effect is apparent in other cell types (62). A group of polypeptides, 40 – 47 kd in size with a major species of 47 kd, purified to 95% homogeneity from HeLa cells, have the DNA binding properties of AP-1 and in addition stimulate transcription of SV40 and MTIIA *in vitro* (115,116). Most viral and cellular genes that respond to AP-1, such as collagenase, stromelysin, MTIIA and SV40, are also induced by treating cells with phorbol ester tumor promoters such as TPA (115,116). In addition, multiple synthetic copies of the consensus AP-1 binding site act as TPA-inducible enhancers in plasmid constructs following transfection into HeLa cells (115,116). Because activation of transcription by TPA occurs by post-translational mechanisms (117), and since TPA activates protein kinase C, it seems likely that AP-1 needs to be phosphorylated in order to increase its transcription activating capability.

6.3.2 The v-jun oncogene is probably derived from the cellular gene encoding AP-1

A role for AP-1 in oncogenic transformation was suggested by the fact that its activity could be modulated by phorbol ester tumor promoters and further evidence for this came from studies on the avian sarcoma virus 17 (ASV17). ASV17 induces fibrosarcomas in chickens and transforms chick embryo fibroblasts into spindle-shaped neoplastic cells (118). Its oncogene, known as v-*jun*, shows structural and functional homology with the yeast transcription factor GCN4 involved in regulating genes required for amino acid biosynthesis (119 – 121). The conserved region is restricted to the C-terminus of v-*jun*, which has 44% homology with the C-terminus of GCN4, known to be the DNA binding domain (120). This suggested that the v-*jun* oncogene was derived from a cellular gene involved in transcriptional control. A possible candiate for this arose when the AP-1 consensus binding site was found to be very similar to that of GCN4, TGACTCA (119). This relationship was confirmed by isolating and sequencing the c-*jun* proto-oncogene, which has greater than 80% homology with v-*jun* sequences. Expression of the c-*jun* cDNA in bacteria produced a protein with sequence-specific DNA binding properties identical to AP-1 (122,123), strongly supporting the idea that AP-1 is encoded by c-*jun*. It seems likely, therefore, that v-*jun* transforms cells by transcriptionally activating cellular genes that have AP-1 binding sites and that a subset, or all, of these may also be induced by phorbol ester tumor promoters through the action of AP-1.

The protein encoded by v-*jun* binds to DNA with the same specificity as AP-1, the c-*jun* gene product, indicating that the mechanism by which the proto-oncogene is activated does not involve any major alteration in DNA binding specificity but more likely a change in the way in which the transcription factor interacts with other proteins (124). Recent data indicate that the GCN4 protein is not the direct yeast equivalent of AP-1. Both *S.cerevisiae* and *Schizosaccharomyces pombe* contain both classical GCN4 and a protein with the SV40 enhancer binding properties of AP-1 (125,126), further supporting the idea that there exists a family of proteins binding to closely related DNA sequences.

6.3.3 AP-1-related proteins are regulated during differentiation

That transcription factors are regulated during differentation is illustrated by studies on murine AP-1-related proteins. The polyoma virus enhancer has an AP-1 motif in the α domain (see later; *Figure 4*), and mutational analysis has shown this domain to be required for transcription and DNA replication (127). The murine equivalent of AP-1, known as PEA1 (128), is regulated during the differentiation of embryonal carcinoma (EC) stem cells such that the binding activity is increased when F9 EC cells differentiate to parietal endoderm (129). It is an interesting speculation that the absence of PEA1 in EC stem cells may account for some of the host range properties of polyoma virus, which is poorly expressed in EC stem cells relative to differentiated murine cells (130,131). The post-translational activation of AP-1 by tumor promoters in HeLa cells is consistent with a mechanism for activating PEA1 during EC cell differentiation.

6.3.4 Sph/octamer motifs

The Sph motif is a repeated sequence consisting of the 9 bp AAG^{T}/$_{C}$ATGCA. It is required for wild-type enhancer activity in a number of cell lines (62,97) and tandem repeats of this sequence have enhancer activity (63). The junction of the direct repeat creates a second motif (ATGCAAAG) strongly homologous to the octamer (ATGCAAAT) found in other promoter and enhancer elements (see discussion on OTF1 and OTF2 in Section 5.3.1), and thus in the SV40 enhancer the Sph and octamer (OCT) motifs overlap (*Figure 2*). Cell-specific effects of these two motifs were shown by studying the expression of mutants in different cell lines. These studies indicate that in HeLa cell extracts proteins bind to the Sph motifs but not to the octamer motif, and that the converse situation occurs in lymphoid cell extracts (132,133). These binding sites correlate with the phenotype of the mutations *in vivo*; the Sph motif but not the octamer motif is active in HeLa cells, whereas in lymphoid cells the octamer is active (62,97).

Earlier we discussed the octamer binding proteins OTF1 and OTF2. Other octamer motif binding activities have been reported, such as the OBP 100 protein purified by octamer affinity chromatography (80) which is probably the same as OTF1. Other activities are less well defined (79).

6.3.5 TCI and TCII motifs

The TCI and TCII motifs bind the HeLa cell transcription factor AP-2 (134) (*Figure 2*). The consensus binding site for AP-2 is CCCCAGGC and again this is not restricted to enhancers but occurs in a number of other *cis* control regions, such as the human metallothionein IIA, human proenkephalin (-80 to -65), human collagenase (-65 to -50), murine H-2K^b (-185 to -163) and adenovirus major late promoters (-135 to -120). Two other transcription factors, NF-$\varkappa$B (135) and H2TF1 (136), bind to related sequences and are discussed later in this chapter.

The purified 52 kd AP-1 polypeptide stimulates transcription from the SV40 early promoter *in vitro* and not only binds to the TCI – TCII motif but also with lower affinity to the SV40 early promoter regions recognized by Sp1 (in the 21 bp repeats) and large T-antigen (over the early transcription initiation site) (134). When AP-2 and Sp1 are incubated together the footprint protection pattern over the 21 bp region is extended, suggesting that AP-2 and Sp1 cooperate (134). In contrast, the sequence-specific binding of AP-2 to the enhancer and early promoter is inhibited by large T-antigen by a mechanism that does not entail binding site competition but rather protein – protein complexing between AP-2 and large T-antigen. Thus, with respect to the 21 bp repeats in the SV40 early control region, AP-2 can have either a positive binding interaction with Sp1 or a negative interaction with large T-antigen.

The AP-2 consensus sequence also mediates transcriptional induction in response to phorbol ester tumor promoters and cAMP elevating agents (137,138). Thus, unlike the AP-1 motif that activates only in response to phorbol esters, the AP-2 sequence mediates transcriptional induction in response to two different signal transduction pathways.

6.3.6 GTI motif

The GTI and GTII motifs are highly homologous to each other (10/12 bases) and to the so-called enhancer 'core sequence' $(GTGG^A/_T{}^A/_T{}^A/_TG)$ (102). A 20 kd protein, EBP20, purified from rat liver binds to the GTI motif in SV40 and related sequences in murine sarcoma virus and polyoma virus (54). As discussed above, EBP20 also binds to the CCAAT motif of HSV tk promoter, so it may be that this molecule has either two DNA binding domains (probably the case for some yeast *trans*-activators, see below) or one domain with degenerate specificity.

Three other proteins bind to the GTI motif, GTIA, GTIB and GTIC (139), the minimal recognition sequence involved being GGGTGTGG, which overlaps with the enhancer 'core' sequence. Of these three proteins, ubiquitous GTIA and GTIB also bind the 21 bp repeats of the early promoter, and GTIA is most probably Sp1, although its binding to the GTI motif is at least 10 times less efficient than to its cognate sequences in the 21 bp repeats; the identity of GTIB remains to be elucidated.

GTIC is cell-specific because it cannot be detected in all cell extracts, such as BJA-B lymphoid cell extracts, and there are clear differences between cell lines, such as HeLa and F9 EC, in which GTIC single point mutations behave differently (139). This same protein binds to an homologous sequence required for activity of the β-globin promoter *in vivo*, so it appears that the protein binding to GTI can function both in enhancers and promoters. This same study showed by competition with the HSV tk or MHC Eα CCAAT motifs that GTIA, B and C are unrelated to EBP20.

6.3.7 GTII motif

The GTII motif is just upstream of the 72 bp repeat and binds at least four different proteins, GTIIA, GTIIBα, GTIIBβ and GTIIC (138). Although the GTII motif has striking homology with the GTI motif, they bind distinct proteins. The four GTII binding proteins recognize three overlapping motifs (*Figure 2*); GTIIBα and GTIIBβ recognize the same binding site. In contrast to other GTII proteins, GTIIC is cell-specific since it is absent from lymphoid cells, and this correlates with the phenotype of GTIIC mutations in such cells.

The GTIIC protein also binds to an enhancer motif in polyoma virus but with greater affinity to that in the polyoma virus mutant F9.1, selected for growth in F9 EC cells. This virus has a point mutation at nucleotide 5233 which creates a GTIIC binding site (TAGAATGT to TGGAATGT), thus possibly explaining the increased expression of this mutant in EC cells (131,140). We discuss this point in greater detail in Section 8.1.

The AP-4 protein also binds to the GTII motif where it acts in concert with AP-1 to activate SV40 late transcription (141).

6.3.8 The purine (Pu) motif

The Pu box was defined by studying selected enhancer deletion mutants with restricted cell-type specificity. When linearized SV40 DNA without the 72 bp repeat sequences was transfected into CV1 cells, virus that grew had undergone a duplication of a DNA segment on the late side of the 72 bp repeats (103,142). This duplicated sequence (379 to 295) possesses enhancer activity in CV1 and lymphoid cells only (142), and the motifs required for each cell-type specificity can be functionally separated into the Pu and KID motifs required for activity in lymphoid cells and kidney epithelial CV1 cells respectively. Proteins recognizing the Pu motif are only detectable in lymphoid cells and bind to a sequence comprising nine purines AAAGAGGAA (142). This Pu box is also present in the enhancer of the lymphotrophic papovavirus and binds the same factor. The relationship of this lymphoid-specific factor to those previously discussed remains to be determined.

The preceding discussion serves to demonstrate that enhancers are composed of sequence motifs, each of which has the potential to bind one or more of a number of different proteins. Just what features determine the protein(s) that bind remain to be determined.

7. Cell-specific gene expression

We now discuss how transcription factors act together to produce cell-specific gene expression, concentrating on two examples, immunoglobulin and albumin gene expression. [The reader is also referred to a detailed review of immunoglobulin gene expression in a companion volume of the *Frontiers in Molecular Biology* series (239).] This is followed by a more general discussion of transcriptional control in embryonic stem cells. We then highlight situations in mammalian development which we anticipate will utilize novel transcription factors and are likely to progress considerably in the near future.

7.1 Immunoglobulin gene expression

The expression of immunoglobulin genes is restricted to cells of the B lineage. In pre-B cells rearrangement and transcription of the heavy chain gene occurs followed in more mature cells by rearrangement and expression of light chain genes. Cell-specific expression requires promoter and enhancer sequences on both the heavy and light chain genes and, as we discussed earlier, the octamer motif is important in mediating this effect (73). The octamer is found in the upstream promoter regions of all heavy and light chain genes (~65 bp in opposite orientations) and in the heavy chain gene enhancer. The octamer motif alone can confer

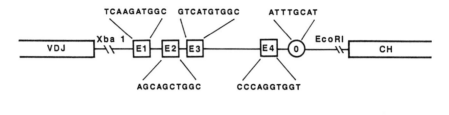

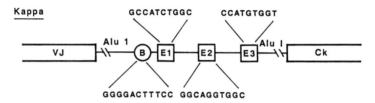

Figure 3.   Protein binding sites in the immunoglobulin heavy chain and kappa light chain enhancers. Each map shows the protein binding sites defined by *in vivo* footprinting, the E motifs (143). The conserved octamer sequence (O) in the heavy chain and the NF-κB binding site (B) in the kappa light chain are indicated. Open boxes indicate exons for variable and constant regions.

lymphoid-specific promoter activity onto a TATA motif (73). Despite this specificity, there are both ubiquitous and lymphoid-specific octamer binding proteins (71,78), and it remains an open question just what features of immunoglobulin promoters determine which protein binds.

In vivo footprinting defined protein binding sites, the E motifs, in the heavy and kappa chain enhancers (see *Figure 3*), the occupancy of which could only be detected in B cells (143). However, *in vitro* analyses of these sites detect the binding activities in most, if not all, cells (144,145). As discussed above, this may be because chromatin is activated in a cell-specific manner *in vivo* so that binding sites are exposed. This step may not be represented *in vitro*, where DNA and not chromatin is used.

Distinct proteins, NF-μE1 and NF-μE3, bind *in vitro* to E1 and E3 motifs in the heavy chain enhancer (144,145). NF-μE3 also binds to E3 in the kappa enhancer, indicating that a single protein can bind to both heavy and light chain enhancers. Both NF-μE1 and NF-μE3 are ubiquitous, as are the two different proteins that bind to the heavy (NF-μE2) and light (NF-κE2) chain E2 motifs. So far no binding to the μE4 and κE1 motifs has been detected (145). *In vivo* analysis of the E motifs has shown that they behave as transcription activating sites, although in the complete heavy chain enhancer no one site is critical, indicating an apparent functional redundancy (145).

The GGGACTTTCC motif present in the kappa chain enhancer binds the protein NF-$\varkappa$B (135). This protein is induced by a post-translational mechanism in pre-B cells stimulated with bacterial lipopolysaccharide (LPS). The combined action of cycloheximide and LPS causes super-induction of NF-$\varkappa$B (135). Phorbol esters also induce an activity similar to NF-$\varkappa$B in B lymphoid and other cells (such as T and HeLa), so its activity is not restricted to B cells. The activation of kappa chain expression that occurs in B cells upon treatment with LPS correlates with an induction of NF-$\varkappa$B, suggesting a role for this factor in transcriptional activation. This is confirmed by *in vivo* studies with the NF-$\varkappa$B sequence motif (146). In fact, NF-$\varkappa$B is undetectable in some kappa chain expressing cells, suggesting that it acts in the initial activation, rather than maintenance, of enhancer function (147). In T cells NF-$\varkappa$B also stimulates transcription of the human immunodeficiency virus genome through an NF-$\varkappa$B binding site (148). The SV40 enhancer also contains an NF-$\varkappa$B binding site overlapping the TCII and GTI motifs (144).

Recently, a cDNA was cloned that encodes part of a protein that binds to a motif similar to that of NF-$\varkappa$B (149). Because at least one other protein, known as H2TF1, binds to a similar motif (136,150), it can only be described at present as possibly encoding NF-$\varkappa$B. The H2TF1 protein binds to the conserved motif TGGGGATTCCCCA located about 160 bp upstream of the H-2K^b class I MHC gene and $\beta 2$ microglobulin enhancers (150). H2TF1 is ubiquitous and present in a wide variety of mammalian cells, in contrast to NF-$\varkappa$B. The H2TF1 motif is also recognized by NF-$\varkappa$B, although with a different affinity; NF-$\varkappa$B binds to both sites with equal affinity, whereas H2TF1 binds with greater affinity to the MHC site relative to the $\varkappa$ enhancer site (151), and they can also be distinguished by major groove contacts (151). Because of these similarities, it is unclear whether NF-$\varkappa$B and H2TF1 represent different modifications of the same protein or are encoded by different genes.

The approach taken to cloning this DNA binding domain was to screen a λ phage expression library with the recognition motif (149). The cDNA isolated hybridizes to a single copy gene in the human genome and is expressed as a 10 kb mRNA in both B and non-B cells, thus strengthening the possibility that NF-$\varkappa$B and H2TF1 are distinct modifications of the same protein.

Our discussion on the expression of immunoglobulin genes emphasizes that cell-specific expression can be effected by combining one motif that binds a cell-specific protein with numerous motifs that bind ubiquitous proteins. This general theme has arisen several times in this chapter so far, such as in the expression of histone H2B gene. It is possible that cell-specific proteins could enable other ubiquitous proteins to bind, and thus are rate-limiting in the assembly of a competent transcription complex.

7.2 Albumin gene expression

The albumin gene is expressed primarily in liver and is coordinately activated with the linked α-fetoprotein gene during fetal development.

The DNA sequences required for transcriptional activation during development include an enhancer located some 10 kb upstream (152) although, once activated, tissue-specific expression is maintained by sequences within 200 bp of the gene (153).

A major advance in studying tissue-specific gene expression occurred when extracts from adult rat liver tissue were shown to be competent for accurate and efficient transcription from the albumin promoter (154). This is specific for this tissue since extracts prepared from spleen and brain transcribe the albumin promoter poorly and sequences defined *in vivo* are required for albumin transcription *in vitro* (154). These sequences, located between −170 and −55, confer liver extract-specific activation onto heterologous promoters, suggesting that they contain binding sites for liver-specific transcription factors. At least six binding sites exist in this region, four of which are occupied by transcription factors enriched in liver (155). One of these is related to NFI and is possibly NF-Y (156). Three of the liver-specific sites bind a similar protein that, from a number of biochemical criteria such as heat stability, is rat liver EBP20, the protein which binds to the SV40 GTI motif and CCAAT box (54). How EBP20 could confer liver-specific expression is unclear, but one distinct possibility would be that a liver-specific post-transcriptional modification makes the protein able to activate transcription in the liver cell environment.

Interestingly, binding of the liver-enriched heat-stable factor hinders the binding of a ubiquitous protein to a neighboring site, although both sites are required for maximum *in vitro* transcription (155,157). One possibility to explain this is that the ubiquitous factor must bind transiently to enable the heat-stable factor to bind, but that once bound, it excludes binding of the ubiquitous factor.

An important step forward in identifying liver-specific transcription factors came with the development of tissue-specific transcription extracts and it should now be possible to study other tissue-specific genes in a similar manner.

8. Transcriptional control in embryonic stem cells

During the past several years much effort has been directed towards identifying proteins that regulate transcription in embryonic stem cells, using in many cases the embryonal carcinoma (EC) cell system. This is because EC cells are thought to be similar to embryonic stem cells in the embryo proper and their use overcomes the problem of the limited material available from the embryo.

Certain well-characterized enhancer and promoter sequences can be used to probe EC cell transcription because some of them, such as the polyoma, Moloney murine leukemia virus (MoMLV) and SV40 enhancers, are differentially active between EC and differentiated cells (158 − 161).

The enhancer effect is generally reduced in EC cells relative to differentiated derivatives and many investigators have attempted to determine why this is so. Two obvious possibilities are that transcription factors are limiting or that transcriptional repressors exist in EC cells. Evidence has been obtained in support of both ideas.

8.1 Transcription factors are limiting in EC cells

That certain transcription factors are limiting in EC cells has been shown by studies on the polyoma virus enhancer. This is located between nucleotides 5021 and 5262 of the viral genome. *In vivo* analysis has shown that, like other enhancers, it consists of a number of modules containing motifs variably shared with other enhancers (127,162,163). The structure of this enhancer together with the functional modules is shown in *Figure 4*.

The left side of the enhancer (5021 to 5128) binds at least two murine proteins, PEA1 and PEA2 (128). As discussed above (Section 6.3.3), PEA1 is most probably the murine equivalent of human AP-1. Although PEA1 is present in differentiated murine cells, it is quantitatively reduced in

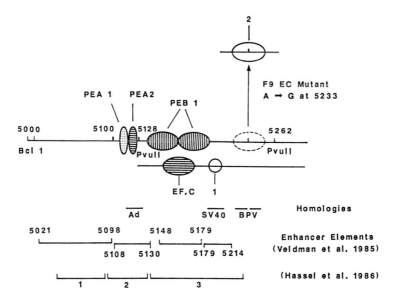

Figure 4. Protein binding sites in the polyoma virus enhancer. The map shows the protein binding sites defined in studies discussed in the text. Homologies to other viral enhancer regions are indicated. Ad, adenovirus E1A enhancer homology; SV40, SV40 core enhancer element homology; BPV, bovine papilloma virus enhancer homology. Also shown are the positions of the functional enhancer elements defined by Veldman *et al.* (127) and Hassel *et al.* (162). At the top of the figure, the protein DNA binding enhanced in the F9 EC mutant F9.1, containing a nucleotide substitution at 5233, is indicated. In the diagram DNA binding proteins (1) and (2) were defined by Fujimura (166) and Kovesdi *et al.* (167) respectively and are discussed in the text.

F9 EC, possibly providing a molecular explanation for the decreased activity in EC cells (129).

Not all enhancer binding factors are reduced in EC cells. This is clearly illustrated by EF.C, which is present in a wide variety of cells, including F9 EC cells, at quantitatively similar levels (164). EF.C binds to an element (5155 to 5174; see *Figure 4*) defined genetically as important for enhancer function *in vivo*. Other proteins that bind to this region have been reported, including PEB1 (5146 to 5196) (165) and another binding between nucleotides 5180 and 5200 (166). This may be another situation where multiple proteins bind to overlapping motifs.

The polyoma virus enhancer functions poorly in EC cells and viruses selected for growth in these cells have mutated enhancer elements (131). Such mutations produce duplicated sequences or point mutations and confer on the enhancer increased activity not only in EC cells but also in differentiated cells (163). The point mutation in the F9.1 polyoma virus produces an A to G transition at nucleotide 5233 (*Figure 4*), enabling this virus to grow with greater efficiency in F9 EC cells (131). This transition creates a protein binding site for a factor that binds with reduced efficiency to the wild-type sequence (167) and, as discussed earlier (Section 6.3.7), is most probably related to GTIIC (138).

Another example of an enhancer that functions inefficiently in EC cells is that of MoMLV (160). Although there is evidence for an EC-specific repressor acting on this enhancer (see below) it also is clear that transcription factors required for enhancer function are limiting, for example, NFI and enhancer core binding proteins (168).

8.2 *Trans*-acting repression in EC cells

Another interesting aspect of the polyoma virus EC mutants concerns their regulation by adenovirus E1A. E1A, the immediate early gene product of adenovirus, has two well-characterized *trans*-acting functions: it not only transcriptionally activates promoters (169), but also represses enhancer-stimulated transcription (170,171). In the case of its effects on the polyoma virus enhancer, the wild-type is repressed but the F9.1 mutation is not (172), indicating that the point mutation also limits the effect of E1A repression. By a number of criteria an E1A-like activity exists in EC cells, one example being that a *trans*-acting transcriptional repressor activity operates *in vivo* on the wild-type polyoma enhancer (173), so perhaps the F9.1 mutation also enables the virus to overcome the repression in EC cells. The SV40 enhancer also has reduced activity in EC cells (158) and again a *trans*-acting repressor exerts an effect (174).

Perhaps the clearest example of repression in EC cells concerns the MoMLV enhancer, which represses promoter activity in *cis* (160), and again a repressor effect operates in *trans* (161). So far, EC cell *trans*-acting repression has only been demonstrated *in vivo*, and it will be a major step forward when these effects are reproduced *in vitro*, thus enabling them to be characterized biochemically.

8.3 EC stem cell transcription activators

As we have discussed, transcription factors such as AP-1 and NFI are limiting in EC cells yet certain transcriptional control sequences function efficiently (175), suggesting that some factors are more abundant in these cells. One of these, called E2F, has been characterized. E2F was originally defined as a transcription factor that accumulates during adenovirus infection. It requires the adenovirus E1A gene product for induction in infected cells (176) and binding site motifs exist in a number of adenovirus promoters (177). The E2F activity is almost undetectable in differentiated cells, such as HeLa, but present at high levels in EC cells (178,179), providing further support for the idea of a cellular E1A-like activity and suggesting a function in EC cell-specific transcriptional activation and perhaps therefore in early embryonic development.

We conclude our discussion of embryonic stem cell transcription by emphasizing that these cells possess novel transcription activities and factors. We assume these to be concerned with their role as stem cells, perhaps arguing that stem cell transcriptional control is fundamentally different from that in differentiated cells.

9. Transcription factors and development

So far we only have hints on transcription factors that function during mammalian development. This is in contrast to the situation in *Drosophila*, where some of the genes that determine the insect body plan encode a functional DNA binding domain, the homeo domain, suggesting that these molecules act at the transcriptional level (for review see 180). In a number of cases the expression of mammalian genes with related domains also shows developmental regulation (181) and by analogy with *Drosophila* these gene products could well function as regulators of transcription.

The developmentally regulated sex determining region of the mammalian Y chromosome may also encode a molecule involved in transcriptional regulation (182). The Y chromosome determines the sex of the mammalian embryo. Cloning of the testis-determining region of this chromosome indicates that part of it encodes a protein with multiple DNA binding fingers (see Section 12), a characteristic of transcription factors such as TFIIIA and Sp1, again suggesting a role in regulating transcription.

Other genes sharing the finger motif have been isolated by searching for homology with finger-containing *Drosophila* segmentation genes, such as *Kruppel* (183,184). In a number of cases their expression is developmentally regulated and the protein products localize to nuclei, against suggesting a role in transcription control (183,185).

10. Lessons from yeast

The analysis of transcriptional control in yeast has advanced somewhat further than in higher eukaryotes, partly due to the smaller genome but also because of the ease with which genetical analysis can be performed. We feel that some of the principles involved in yeast transcriptional control will turn out to have analogies with situations in higher eukaryotes and therefore now discuss a number of examples. It is noteworthy that these analogies could involve functional convergence since yeast transcriptional control sequences have recently been shown to function efficiently in mammalian cells (186,187).

10.1 Upstream activation sites are binding sites for regulatory proteins

Transcriptional control sequences of yeast genes usually have three essential control elements: the TATA box, the initiation site and at least one upstream activation site (UAS). The UAS element is found upstream from the TATA box and can be inverted or moved some distance relative to it (188,189), thus manifesting properties in common with enhancer sequences of higher eukaryotes. In many cases these UAS sites mediate gene-specific regulation, for example, the response to galactose leading to the transcriptional activation of genes encoding galactose catabolizing enzymes (188). Regulatory mutations have been used to define genes the products of which bind to UAS sites to confer transcriptional activation, examples being GAL4, GCN4 and the HAP series of proteins. In these cases purified protein binds directly to the relevant UAS, with distinct protein domains required for DNA binding and transcriptional activation (190 – 194).

10.1.1 HAP1, -2 and -3

The *HAP1, -2* and *-3* loci encode proteins required for the transcriptional activation of a number of genes, such as the *CYC1* gene encoding iso-1-cytochrome *c*. *CYC1* has two adjacent UASs, UAS1 and UAS2, activated by either HAP1 or the combined action of HAP2 and HAP3, respectively (195 – 197). UAS1 can be further divided into two regions, A and B, both required for full activity. Another yeast protein, RC2, binds in UAS1 to the same sequence as HAP1 but the binding of the two proteins is distinguished by minor differences in DNA contacts and is mutually exclusive (195). HAP1 not only binds to the *CYC1* UAS but also to the *CYC7* UAS (196), although these two sites share no sequence similarity, and indirect evidence suggests that HAP1 has only one DNA binding domain (196). In summary, the HAP1 protein binds a sequence shared with another activator protein and possesses a DNA binding property that enables recognition of distinct binding sites. Earlier in this chapter we discussed a number of higher eukaryotic proteins that could operate in a similar fashion.

The *CYC1* UAS2 binds proteins encoded by *HAP2* and *HAP3*, and in this case its full activity is dependent on HAP2 and HAP3 binding together, that is, they are required to bind interdependently (197,198). It is noteworthy that the UAS2 binding site is homologous to the CCAAT box, and we already know of a number of proteins that bind this motif in higher eukaryotes. It would not be surprising if multiple proteins bind simultaneously to this sequence in higher eukaryotes. Recent evidence supports this idea (199). Surprisingly, the subunits of the yeast HAP2 – HAP3 complex and a human CCAAT-binding protein are also functionally interchangeable (200).

10.2 The same protein can activate and repress transcription

Mating type information in yeast is repressed in *cis* by sequences referred to as silencers, composed of distinct regulatory modules that bind defined *trans*-acting proteins (201). Silencers have a number of characteristics in common with higher eukaryotic enhancers, for example, orientation- and distance-independent effects, but they repress rather than activate transcription. One of the proteins required for silencer function, called RAP1, not only binds to sequences within the silencer but also to UAS elements of other genes where it functions in activation (201,202), suggesting that it can act as both repressor and activator (202). These two opposite activities of the same protein may depend on the context of the binding site and again similar situations may exist in higher eukaryotes.

10.3 Transcriptional activation requires an amphipathic helix

A number of yeast regulatory proteins, for example GAL4 (190), GCN4 (192) and HAP1 (195), bind to their cognate UASs in the absence of additional yeast proteins and can be divided into separate domains required for DNA binding and transcriptional activation. The GAL4 protein has been characterized extensively; residues 1 to 147 have the capacity to bind DNA *in vivo* and *in vitro* but fail to activate transcription (190). The transcription activating domain was identified by studies on GAL4 fusion proteins, bearing the DNA binding domain of the bacterial repressor LexA in place of the GAL4 domain, which can activate transcription by binding to a LexA operator positioned upstream of a yeast gene (193). Through this type of experiment two transcription activating domains were identified, each of which is highly acidic and has no homology with other activators such as GCN4 (190). In fact, the activating domain can be replaced by random regions of the *E.coli* genome fused to the GAL4 DNA binding domain and again the only characteristic that these genomic fragments have in common is the acidic nature of the encoded polypeptide (203). Point mutations that increase the activation efficiency make the sequence more acidic whereas some of those that decrease the activation capacity decrease the acidity of the sequence (204). A totally synthetic activating region has been described consisting of a 15-amino-acid acidic peptide in an amphipathic helix, that is, one with a hydrophilic face bearing acidic residues, the other face being hydrophobic (205). A peptide with the same amino acid content but presented in a scrambled order lacks any transcription activating function (205). The reason for this requirement for an amphipathic helix is unclear, but it is also found in the GCN4 activating region (192). It may be necessary for the protein to contact other proteins, such as RNA polymerase, although this remains speculative. It will no doubt be a structure to expect when protein sequences for higher eukaryotic *trans*-activators become available. That GAL4 can function as a transcriptional activator in mammalian cells (186,187) supports this idea.

11. RNA polymerase I

Eukaryotic ribosomal RNA (rRNA) genes are arranged in tandem arrays and transcribed by RNA polymerase I. This transcription occurs in nucleoli and constitutes about one-half of the total transcription in the cell. It is regulated in response to a wide variety of stimuli such as growth rate, viral infection and heat shock, and consequently has been studied by many workers.

Ribosomal RNA genes have one type of promoter, repeated about 200 times per haploid genome, consisting of two domains. One of these is absolutely required for promoter activity and includes sequences from the initiation site to about −50; the other, with variable quantitative effects, is located in the −50 to −150 region (206,207) and is both distance- and orientation-dependent. These promoter sequences have superimposed terminator sequences. RNA polymerase I initiation may even be enhanced by transcription termination sequences, suggestive of a multi-functional protein (208−210; see also Chapter 3).

An interesting property of rRNA transcription is the species specificity of promoter recognition: the RNA polymerase I transcriptional machinery of one species, such as man, will not recognize the promoter of another species such as mouse. This specificity is mimicked *in vitro* in extracts prepared from different species and is reflected in the degree to which different species' promoters have diverged, in contrast to the coding sequences (211,212).

Transcription factors that act together with RNA polymerase I have been partially characterized in a number of studies (213–215). One of these, called SL1, isolated from human cells, confers promoter selectivity in a mouse cell extract onto a human rRNA gene and thus is a polymerase I promoter species specificity factor. Antibodies against SL1 stain nucleoli and inhibit SL1-dependent transcription *in vitro* (213). This protein does not have an intrinsic DNA binding activity but rather binds to promoter sequences in cooperation with another protein (215).

Studies on *Acanthamoeba* suggest that the activity of RNA polymerase I can be modified directly (216) because the enzyme from encysted cells is unable to utilize the promoter, in contrast to that from growing cells. This is not due to the inactivation of any accessory proteins which are present, suggesting that modification of the enzyme is also an important level of regulation, possibly by controlling contact with promoter factors (217).

In summary, the regulation of RNA polymerase I transcription has a number of properties in common with regulation of RNA polymerase II such as the requirement for *trans*-acting protein factors that bind to upstream promoter sequences. As they become biochemically characterized it will be most interesting to determine what unique features of these molecules determine their interaction(s) with the RNA polymerase I system.

12. RNA polymerase III

RNA polymerase III transcribes genes encoding a variety of small, stable RNAs, for example tRNA, 5S rRNA and some snRNAs. Promoter sequences for RNA polymerase III, originally defined within the coding sequence, were subsequently shown also to reside outside the coding sequence, where they may act alone or together with promoter domains in the coding sequence (218).

Chromatographic separation of crude nuclear extracts indicates that at least three separate fractions are required for polymerase III transcription (219,220). One of these, TFIIIA, is required only for 5S gene transcription and binds in a sequence-specific fashion to the internal promoter of 5S genes in the absence of other proteins (221). TFIIIA also binds to the 5S RNA, stabilizing it in the 7S storage particle (222). The 38 kd TFIIIA protein is composed of two separable functional domains, one responsible for DNA binding and the other for protein–protein interactions with other components of the transcriptional apparatus (223,224). A cDNA encoding TFIIIA has been isolated and sequence analysis shows that the DNA binding domain consists of tandemly repeated units of 28–30 amino acids, each containing two cysteines and two histidines at invariant positions (225). It has been proposed that each of these units folds as a single domain, centered on a zinc ion coordinated by histidine and cysteine, that binds to five nucleotides in the DNA (226,227). These units have been referred to as zinc 'fingers' and are present in a number of other DNA binding proteins, such as Sp1, and potential regulatory molecules, such as TDF and the *Kruppel* gene product (228).

The two other fractions, TFIIIB and TFIIIC, are also involved in formation of a stable pre-initiation complex on 5S genes (219,220). The formation of the TFIIIA DNA complex is required before TFIIIC can bind, followed by the subsequent stable association of TFIIIB (220). Other class III genes, such as tRNA genes and the adenovirus VA gene, do not utilize TFIIIA but instead form a stable complex with TFIIIC and TFIIIB (220,229). Like TFIIIA, the TFIIIC fraction binds to DNA but is composed of two separable activities, TFIIIC1 and TFIIIC2, that bind to different domains of the promoter (230). In adenovirus VA gene promoters, which are composed of two functional domains, TFIIIC1 binds to the A box and TFIIIC2 to the B box. There is some cooperation between these two factors because TFIIIC1 binds poorly to the A box in the absence of TFIIIC2.

Although most class III genes have intragenic promoter sequences, a number have promoter domains located external to the coding region, where they may act together with intragenic control sequences, as in the 7SL gene (231), or have no requirement for intragenic sequences, as with the 7SK and U6 genes (232–236). Such external promoter domains contain, in the cases so far analyzed, sequence motifs characteristic of RNA-polymerase-II-dependent promoters, for example, TATA-like and octamer motifs (234,235). Point mutations in the TATA motif of the 7SK gene reduce its ability to be transcribed by RNA polymerase III, and both U6 and 7SK genes have two copies of the octamer motif in the upstream region which again are required for optimal stimulation by RNA polymerase III (234–236). These observations suggest that RNA polymerase II and III utilize the same *trans*-acting factors, a not altogether surprising observation since these two enzymes share common subunits.

13. Summary and future perspectives

Regulation of transcriptional initiation in eukaryotic cells is mediated by *trans*-acting protein factors that bind to DNA sequence motifs. These motifs combine to produce activities that can be divided broadly into promoters or enhancers, individual motifs having a broad or restricted distribution. This combinatorial model argues against the idea of one specific transcription factor for one specific gene but, rather, suggests that a limited number of motifs combine in different orders and numbers to produce the temporal and quantitative characteristics of transcriptional activation.

Different *trans*-acting factors can bind to the same motif, clearly illustrated by the CCAAT box and octamer binding factors. What features determine exactly which factor binds to which sequence inside the nucleus are yet to be resolved, but we can imagine that cooperative interactions between proteins bound to surrounding motifs or steric constraints will be important. Alternatively, transcriptional activation through a particular motif may require multiple factors to bind together, a situation found in HAP2/HAP3 *trans*-activation in yeast (195).

It is noteworthy that cell-specific expression in a number of cases is mediated by cell-specific factors binding to common motifs, as seen for the lymphoid-specific octamer binding factor (73,78), and so this may turn out to be a general principle in cell-specific transcription. An alternative approach is to activate ubiquitous factors in a cell-specific fashion, perhaps post-translationally, but we await firm evidence for this.

Extensive functional analysis of higher eukaryotic transcription factors has yet to be performed, but we can anticipate that this task will be aided by generating cDNA clones and antibody probes. Perhaps these transcription factors use the same rules as in yeast, where discrete domains of the molecule have a defined function. This will in turn enable the mechanism of transcriptional activation to be deciphered and allow a distinction between possible mechanisms, such a signals being transmitted through the DNA molecule or transcription factors contacting polymerase subunits. Some progress in this area has been made in yeast (237). Proteins that are capable of activating both transcription and DNA replication, such as NFI and NFIII/OTF1, can also be explored in this way. Are different domains of the protein involved, or do they provide the same function in both situations?

It is important to determine at what level transcription factors are themselves regulated. A number of studies suggest that this can be post-translational (117), implying that the molecule exists in an inactive state beforehand. This situation may exist for rapid transcriptional responses. There are probably other levels of regulation also, perhaps during embryonic development, where gene expression is temporally and spatially controlled. Again, studying the cellular distribution of their mRNAs and the transcription factors themselves will resolve these issues.

In the introduction we suggested that the activation of gene expression should be looked upon as being a two-step process, the first involving decondensation and the second the binding of *trans*-acting factors to *cis*-acting regulatory motifs. We have discussed the factors involved in the second step because the mechanisms involved in the first remain to be determined. The dominant control region (DCR) identified upstream of the human globin gene may be a DNA sequence involved in mediating this first step, perhaps by binding to the nucleoskeleton (238). We need now to address the proteins that bind to DCR-like regions, predicting perhaps that they will turn out to be sequence-specific proteins with activities enabling decondensation, such as topoisomerases.

14. Acknowledgements

N.B.L.T. is a Jenner Fellow of the Lister Institute for Preventive Medicine.

15. References

1. Sollner-Webb,B. and Tower,J. (1986) Transcription of cloned eukaryotic ribosomal RNA genes. *Annu. Rev. Biochem.,* **55**, 801.
2. Ciliberto,G., Cactagnoli,L. and Cortese,R. (1983) Transcription by RNA polymerase III. *Curr. Topics Dev. Biol.,* **18**, 59.
3. Yamamoto,K.R. (1985) Steroid receptor regulated transcription of specific genes and gene networks. *Annu. Rev. Genet.,* **19**, 209.
4. Gasser,S.M. and Laemmli,U.K. (1986) Cohabitation of scaffold binding regions with upstream/enhancer elements of three developmentally regulated genes of *D. melanogaster. Cell,* **46**, 521.
5. Cockerhill,P.N. and Gerrard,W.T. (1986) Chromosomal loop anchorage of the kappa Ig gene locus next to the enhancer in a region containing topoisomerase II sites. *Cell,* **44**, 273.
6. Mann,C., Buhler,J.-M., Treich,I. and Sentenac,A. (1987) RPC 40, a unique gene for a subunit shared between yeast RNA polymerases A and C. *Cell,* **48**, 627.
7. Allison,L.A., Moyle,M., Shales,M. and Ingles,C.J. (1985) Extensive homology among the largest subunits of eukaryotic and prokaryotic RNA polymerases. *Cell,* **42**, 599.
8. Biggs,J., Searles,L.L. and Greenleaf,A.L. (1985) Structure of the eukaryotic transcription apparatus; features of the gene for the largest subunit of *Drosophila* RNA polymerase II. *Cell,* **42**, 611.
9. Buhler,J.-M., Riva,M., Mann,C., Thuriaux,P., Memet,S., Micouin,J.Y., Treich,I., Mariotte,S. and Sentenac,A. (1987) Eukaryotic RNA polymerases, subunits and genes. In *RNA Polymerase and the Regulation of Transcription.* Reznikoff,W.S., Burgess,R.R., Dahlberg,J.E., Gross,C.A., Record,M.T. and Wickens,M.P. (eds), Elsevier, New York, p. 25.
10. Sweetser,D., Nunet,M. and Young,R.A. (1987) Prokaryotic and eukaryotic RNA polymerases have homologous core subunits. *Proc. Natl. Acad. Sci. USA,* **84**, 1192.
11. Allison,L.A., Wong,J.K.-C., Fitzpatrick,V.D., Moyle,M. and Ingles,C.J. (1988) The C-terminal domain of the largest subunit of RNA polymerase II of *Saccharomyces cerevisiae, Drosophila melanogaster,* and mammals: a conserved structure with an essential function. *Mol. Cell. Biol.,* **8**, 321.
12. Nonet,M., Sweetser,D. and Young,R.A. (1987) Functional redundancy and structural polymorphism in the large subunit of RNA polymerase II. *Cell,* **50**, 909.
13. Bartolomei,M.S., Halden,N.F., Cullen,C.R. and Corden,J.L. (1988) Genetic analysis of the repetitive carboxyl-terminal domain of the largest subunit of mouse RNA polymerase II. *Mol. Cell. Biol.,* **8**, 330.
14. Matsui,T., Segall,J., Weil,P.A. and Roeder,R.G. (1980) Multiple factors required for accurate initiation of transcription by purified RNA polymerase II. *J. Biol. Chem.,* **244**, 11992.
15. Bird,A.P. (1986) CpG-rich islands and the functions of DNA methylation. *Nature,* **321**, 209.
16. Sawadogo,M. and Roeder,R.G. (1985) Factors involved in specific transcription by human RNA polymerase II: analysis by a rapid and quantitative *in vitro* assay. *Proc. Natl. Acad. Sci. USA,* **82**, 4394.
17. Sawadogo,M. and Roeder,R.G. (1985) Interaction of a gene-specific transcription factor with the adenovirus major late promoter upstream of the TATA box region. *Cell,* **43**, 165.
18. Garcia,J., Wu,F. and Gaynor,R. (1987) Upstream regulatory regions required to stabilise binding to the TATA sequence in an adenovirus early promoter. *Nucleic Acids Res.,* **15**, 8367.

19. Workman,J.L. and Roeder,R. (1987) Binding of transcription factor TFIID to the major late promoter during *in vitro* nucleosome assembly potentiates subsequent initiation by RNA polymerase II. *Cell,* **51**, 613.
20. Davison,B.L., Egley,J.M., Mulvihill,E.R. and Chambon,P. (1983) Formation of stable preinitiation complexes between eukaryotic class B transcription factors and promoter sequences. *Nature,* **301**, 680.
21. Fire,A., Samuels,M. and Sharp,P.A. (1984) Interactions between RNA polymerase II, factors, and template leading to accurate transcription. *J. Biol. Chem.,* **259**, 2509.
22. Hawley,D.K. and Roeder,R.G. (1985) Separation and partial characterization of three functional steps in transcription initiation by human RNA polymerase II. *J. Biol. Chem.,* **260**, 8163.
23. Hawley,D.K. and Roeder,R.G. (1987) Functional steps in transcription initiation and reinitiation from the major late promoter in a HeLa nuclear extract. *J. Biol. Chem.,* **262**, 3452.
24. Reinberg,D., Horikoshi,M. and Roeder,R.G. (1987) Factors involved in specific transcription by mammalian RNA polymerase II: functional analysis of initiation factors TFIIA and TFIID and identification of a new factor operating at sequences downstream of the initiation site. *J. Biol. Chem.,* **262**, 3322.
25. Egly,J.M., Miyamoto,N.G., Moncollin,V. and Chambon,P. (1984) Is actin a transcription initiation factor for RNA polymerase B? *EMBO J.,* **3**, 2363.
26. Samuels,M. and Sharp,P.A. (1986) Purification and characterization of a specific RNA polymerase II transcription factor. *J. Biol. Chem.,* **261**, 2003.
27. Reinberg,D. and Roeder,R.G. (1987) Factors involved in specific transcription by mammalian RNA polymerase II: purification and functional analysis of initiation factors TFIIB and TFIIE. *J. Biol. Chem.,* **262**, 3310.
28. Zheng,X.-M., Moncollin,V., Egly,J.-M. and Chambon,P. (1987) A general transcription factor forms a stable complex with RNA polymerase B (II). *Cell,* **50**, 361.
29. Burton,Z.F., Ortolan,L.G. and Greenblatt,J. (1986) Proteins that bind to RNA polymerase II are required for accurate initiation of transcription at the adenovirus 2 major late promoter. *EMBO J.,* **5**, 2923.
30. Sekimizu,K., Kubo,Y., Segawa,K. and Natori,S. (1981) Difference in phosphorylation of two factors stimulating RNA polymerase II of Ehrlich ascites tumor cells. *Biochemistry,* **20**, 2286.
31. Horikoshi,M., Sekimizu,K., Hirashima,S., Mitsuhashi,Y. and Natori,S. (1985) Structural relationships of the three stimulatory factors of RNA polymerase II from Ehrlich ascites tumor cells. *J. Biol. Chem.,* **260**, 5739.
32. Reinberg,D. and Roeder,R.G. (1987) Factors involved in specific transcription by mammalian RNA polymerase II: transcription factor IIS stimulates elongation of RNA chains. *J. Biol. Chem.,* **262**, 3331.
33. Sagami,I., Tsai,S.Y., Wang,H., Tsai,M.-J. and O'Malley,B.W. (1986) Identification of two factors required for transcription of the ovalbumin gene. *Mol. Cell. Biol.,* **6**, 4259.
34. Tsai,S.Y., Sagami,I., Wang,L.-H., Tsai,M.-J. and O'Malley,B.W. (1987) Interactions between a DNA-binding transcription factor (COUP) and a non-DNA binding factor (S300-II). *Cell,* **50**, 701.
35. Dynan,W.S. and Tjian,R. (1983) The promoter-specific transcription factor Sp1 binds to upstream sequences in the SV40 early promoter. *Cell,* **35**, 79.
36. Dynan,W.S. and Tjian,R. (1983) Isolation of transcription factors that discriminate between different promoters recognized by RNA polymerase II. *Cell,* **32**, 669.
37. Gidoni,D., Kadonaga,J.T., Barrera-Saldana,H., Takahashi,K., Chambon,P. and Tjian,R. (1985) Bidirectional SV40 transcription mediated by tandem Sp1 binding interactions. *Science,* **230**, 511.
38. Gidoni,D., Dynan,W.S. and Tjian,R. (1984) Multiple specific contacts between a mammalian transcription factor and its cognate promoters. *Nature,* **312**, 409.
39. Briggs,M.R., Kadonaga,J.T., Bell,S.P. and Tjian,R. (1986) Purification and biochemical characterization of the promoter-specific transcription factor Sp1. *Science,* **234**, 47.
40. Kadonaga,J.T. and Tjian,R. (1986) Affinity purification of sequence-specific DNA binding proteins. *Proc. Natl. Acad. Sci. USA,* **83**, 5889.
41. Jones,K.A., Yamamoto,K.R. and Tjian,R. (1985) Two distinct transcription factors bind to the HSV thymidine kinase promoter *in vitro*. *Cell,* **42**, 559.
42. Dynan,W.S., Sazer,S., Tjian,R. and Schimke,R.T. (1985) Transcription factor Sp1

recognises a DNA sequence in the mouse dihydrofolate reductase promoter. *Nature*, **319**, 246.

43. Dynan,W.S. and Tjian,R. (1985) Control of eukaryotic messenger RNA synthesis by sequence-specific DNA-binding proteins. *Nature, * **316**, 774.

44. Kadonaga,J.T., Jones,K.A. and Tjian,R. (1986) Promoter-specific activation of RNA polymerase II transcription by Sp1. *Trends Biochem.,* **11**, 20.

45. Jones,K.A. and Tjian,R. (1985) Sp1 binds to promoter sequences and activates herpes simplex virus 'immediate early' gene transcription *in vitro. Nature,* **317**, 197.

46. Kadonaga,J.T., Carner,K.R., Masiarz,F.R. and Tjian,R. (1987) Isolation of cDNA encoding transcription factor Sp1 and functional analysis of the DNA binding domain. *Cell,* **51**, 1079.

47. Bienz,M. (1987) A CCAAT box confers cell-type-specific regulation on the *Xenopus* hsp70 gene in oocytes. *Cell,* **46**, 1037.

48. Morgan,W.D., Williams,G.T., Morimoto,R.I., Green,J., Kingston,R.E. and Tjian,R. (1987) Two transcriptional activators, CCAAT-box-binding transcription factor and heat shock transcription factor, interact with a human hsp70 gene promoter. *Mol. Cell. Biol.,* **7**, 1129.

49. Jones,K.A., Kadonaga,J.T., Rosenfeld,P.J., Kelly,T.J. and Tjian,R. (1987) A cellular DNA-binding protein that activates eukaryotic transcription and DNA replication. *Cell,* **48**, 79.

50. Rosenfeld,P.J. and Kelly,T.J. (1986) Purification of nuclear factor I by DNA recognition site affinity chromatography. *J. Biol. Chem.,* **281**, 1398.

51. Rosenfeld,P.J., O'Neill,E.A., Wides,R.J. and Kelly,T.J. (1987) Sequence-specific interactions between cellular DNA-binding proteins and the adenovirus origin of DNA replication. *Mol. Cell. Biol.,* **7**, 875.

52. Pruijn,G.J.M., van Driel,W. and van der Vliet,P.C. (1986) Nuclear factor III, a novel sequence-specific DNA-binding protein from HeLa cells stimulating adenovirus DNA replication. *Nature, * **322**, 656.

53. Graves,B.J., Johnson,P.F. and McKnight,S.L. (1987) Homologous recognition of a promoter domain common to the MSV LTR and the HSV tk gene. *Cell,* **44**, 565.

54. Johnson,P.F., Landschulz,W.H., Graves,B.J. and McKnight,S.L. (1987) Identification of a rat liver nuclear protein that binds to the enhancer core element of three animal viruses. *Genes Dev.,* **1**, 133.

55. Dorn,A., Durand,B., Marfing,C., Le Meur,M., Benoist,C. and Mathis,D. (1987) The conserved MHC class II boxes X and Y are transcriptional control elements and specifically bind nuclear proteins. *Proc. Natl. Acad. Sci. USA,* **84**, 6249.

56. Dorn,A., Bollekens,J., Staub,A., Benoist,C. and Mathis,D. (1987) A multiplicity of CCAAT box-binding proteins. *Cell,* **50**, 863.

57. Hoofte van Huiysduijnen,R.A.M., Bollekens,J., Dorn,A., Benoist,C. and Mathis,D. (1987) Properties of a CCAAT box binding protein. *Nucleic Acids Res.,* **15**, 7265.

58. Cohen,R.B., Shettery,M. and Kim,C.G. (1986) Partial purification of a nuclear protein that binds to the CCAAT box of the mouse α_1-globin gene. *Mol. Cell. Biol.,* **6**, 821.

59. Kemler,I. and Busslinger,M. (1986) Characterization of two nonallelic pairs of late histone H2A and H2B genes of the sea urchin: differential regulation in the embryo and tissue-enhancer expression in the adult. *Mol. Cell. Biol.,* **6**, 3746.

60. Barberis,A., Superti-Furga,G. and Busslinger,M. (1987) Mutually exclusive interaction of the CCAAT-binding factor and of a displacement protein with overlapping sequences of a histone gene promoter. *Cell,* **50**, 347.

61. Harvey,R.P., Robins,A.J. and Wells,J.R.E. (1982) Independently evolving chicken histone H2B genes: identification of a ubiquitous H2B-specific 5' element. *Nucleic Acids Res.,* **10**, 7851.

62. Nomiyama,H., Fromental,C., Xiao,J.H. and Chambon,P. (1987) Cell-specific activity of the constituent elements of the simian virus 40 enhancer. *Proc. Natl. Acad. Sci. USA,* **84**, 7881.

63. Ondek,B., Shepard,A. and Herr,W. (1987) Discrete elements within the SV40 enhancer region display different cell-specific enhancer activities. *EMBO J.,* **6**, 1017.

64. La Bella,F., Sive,H.L., Roeder,R.G. and Heintz,N. (1988) Cell-cycle regulation of a human histone H2b gene is mediated by the H2b subtype-specific consensus sequence. *Genes Dev.,* **2**, 32.

65. Sive,H.L. and Roeder,R.G. (1986) Interaction of a common factor with conserved promoter and enhancer sequences in histone H2B, immunoglobulin and U2 small nuclear RNA (snRNA) genes. *Proc. Natl. Acad. Sci. USA,* **83**, 6382.
66. Bohmann,D., Keller,W., Dale,T., Scholer,H.R., Tebb,G. and Mattaj,I.W. (1987) A transcription factor which binds to the enhancers of SV40, immunoglobulin heavy chain and U2 snRNA genes. *Nature,* **325**, 268.
67. Preston,C.M., Frame,M.C. and Campbell,M.E.M. (1988) A complex formed between cell components and an HSV structural polypeptide binds to a viral immediate early gene regulatory DNA sequence. *Cell,* **52**, 425.
68. O'Hare,P. and Goding,C.R. (1988) Herpes simplex virus regulatory elements and the immunoglobulin octamer domain bind a common factor and are both targets for virion transactivation. *Cell,* **52**, 435.
69. Sive,H.L., Heintz,N. and Roeder,R.G. (1986) Multiple sequence elements are required for maximal *in vitro* transcription of a human histone H2B gene. *Mol. Cell. Biol.,* **6**, 3329.
70. Heintz,N. and Roeder,R.G. (1984) Transcription of human histone genes in extracts from synchronized HeLa cells. *Proc. Natl. Acad. Sci. USA,* **81**, 1713.
71. Fletcher,C., Heintz,N. and Roeder,R.G. (1987) Purification and characterization of OTF-1, a transcription factor regulating cell cycle expression of a human histone H2b gene. *Cell,* **51**, 773.
72. Pruijn,G., van Driel,J.M., van Milkenberg,R.T. and van der Vliet,P.C. (1987) Promoter and enhancer elements containing a conserved sequence motif are recognised by nuclear factor III, a protein stimulating adenovirus DNA replication. *EMBO J.,* **6**, 3771.
73. Wirth,T., Staudt,L. and Baltimore,D. (1987) An octamer oligonucleotide upstream of a TATA motif is sufficient for lymphoid specific promoter activity. *Nature,* **329**, 174.
74. Falkner,F.G. and Zachua,H.G. (1984) Correct transcription of an immunoglobulin *x* gene requires an upstream fragment containing conserved sequence elements. *Nature,* **310**, 71.
75. Mizushima-Sugano,J. and Roeder,R.G. (1986) Cell-type-specific transcription of an immunoglobulin k light chain gene *in vitro. Proc. Natl. Acad. Sci. USA,* **83**, 8511.
76. Singh,H., Sen,R., Baltimore,D. and Sharp,P.A. (1986) A nuclear factor that binds to a conserved sequence motif in transcriptional control elements of immunoglobulin genes. *Nature,* **319**, 154.
77. Staudt,L.M., Singh,H., Sen,R., Wirth,T., Sharp,P.A. and Baltimore,D. (1986) A lymphoid-specific protein binding to the octamer motif of immunoglobulin genes. *Nature,* **323**, 640.
78. Scheidereit,C., Heguy,A. and Roeder,R. (1987) Identification and purification of a human lymphoid-specific octamer binding protein (OTF-2) that activates transcription of an immunoglobulin promoter *in vitro. Cell,* **51**, 783.
79. Rosales,R., Vigneron,M., Macchi,M., Davidson,I., Xiao,J.H. and Chambon,P. (1987) *In vitro* binding of cell-specific and ubiquitous nuclear proteins to the octamer motif of the SV40 enhancer and related motifs present in other promoters. *EMBO J.,* **6**, 3015.
80. Sturm,R., Baumruker,T., Franza,B.R. and Herr,W. (1987) A 100 kd HeLa cell octamer binding protein (OBP 100) interacts differently with two separate octamer-related sequences within the SV40 enhancer. *Genes Dev.,* **1**, 1147.
81. Manley,J., Fire,A., Cano,A., Sharp,P.A. and Gefter,M. (1980) DNA-dependent transcription of adenovirus genes in a soluble whole-cell extract. *Proc. Natl. Acad. Sci. USA,* **77**, 3855.
82. Hen,R., Sassone-Corsi,P., Corden,J., Gaub,M.P. and Chambon,P. (1982) Sequences upstream from the T-A-T-A box are required *in vivo* and *in vitro* for efficient transcription from the adenovirus serotype 2 major late promoter. *Proc. Natl. Acad. Sci. USA,* **79**, 7132.
83. Carthew,R.W., Chodosh,L.A. and Sharp,P.A. (1985) An RNA polymerase II transcription factor binds to an upstream element in the adenovirus major late promoter. *Cell,* **43**, 439.
84. Chodosh,L.A., Carthew,R.W. and Sharp,P.A. (1986) A single polypeptide possesses the binding and transcription activities of the adenovirus major late transcription factor. *Mol. Cell. Biol.,* **6**, 4723.

85. Chodosh,L.A., Carthew,R.W., Morgan,J.G., Crabtree,G.R. and Sharp,P.A. (1987) The adenovirus major late transcription factor activates the rat γ-fibrinogen promoter. *Science*, **238**, 684.
86. Carthew,R.W., Chodosh,L.A. and Sharp,P.A. (1987) The major late transcription factor binds to and activates the mouse metallothionein 1 promoter. *Genes Dev.*, **1**, 973.
87. Pelham,H.R.B. (1985) Activation of heat-shock genes in eukaryotes. *Trends Genet.*, **1**, 31.
88. Bienz,M. and Pelham,H.R.B. (1986) Heat shock regulatory elements function as an inducible enhancer in the *Xenopus* hsp70 gene and when linked to a heterologous promoter. *Cell*, **45**, 753.
89. Widderrecht,G., Shuey,D.J., Kibbe,W.A. and Parker,C.S. (1987) The *Saccharomyces* and *Drosophila* heat shock transcription factors are identical in size and DNA binding properties. *Cell*, **48**, 507.
90. Sorger,P.K. and Pelham,H.R.B. (1987) Purification and characterization of a heat shock element binding protein from yeast. *EMBO J.*, **6**, 3035.
91. Sorger,P.K., Lewis,M.J. and Pelham,H.R.B. (1987) Heat shock factor is regulated differently in yeast and HeLa cells. *Nature*, **329**, 81.
92. Wu,C. (1984) Two protein-binding sites in chromatin implicated in the activation of heat shock genes. *Nature*, **309**, 229.
93. Shuey,D.J. and Parker,C.S. (1986) Bending of promoter DNA on binding of a heat shock transcription factor. *Nature*, **323**, 459.
94. Montminy,M.R., Sevarino,K.A., Wagner,J.A., Mandel,G. and Goodman,R.H. (1986) Identification of a cAMP-responsive element within the rat somatostatin gene. *Proc. Natl. Acad. Sci. USA*, **83**, 6682.
95. Montminy,M.R. and Bilezikjian,L.M. (1987) Binding of a nuclear protein to the cAMP responsive element of the somatostatin gene. *Nature*, **328**, 175.
96. Lee,K.A.W., Hai,T.-Y., Siva Raman,L., Thimmappaya,B., Hurst,H.C., Jones,N.C. and Green,M. (1987) A cellular protein, activating transcription factor, activates transcription of multiple E1A-inducible adenovirus early promoters. *Proc. Natl. Acad. Sci. USA*, **84**, 8355.
97. Zenke,M., Grundstrom,T., Matthes,H., Wintzerith,M., Schatz,C., Wildeman,A. and Chambon,P. (1986) Multiple sequence motifs are involved in SV40 enhancer function. *EMBO J.*, **5**, 387.
98. Herr,W. and Gluzman,Y. (1985) Duplications of a mutated simian virus 40 enhancer restore its activity. *Nature*, **313**, 711.
99. Herr,W. and Clarke,J. (1986) The SV40 enhancer is composed of multiple functional elements that can compensate for one another. *Cell*, **45**, 461.
100. Clarke,J. and Herr,W. (1987) Activation of mutated simian virus 40 enhancers by amplification of wild-type enhancer elements. *J. Virol.*, **61**, 3536.
101. Serfling,E., Jasin,M. and Schaffner,W. (1985) Enhancers and eukaryotic gene transcription. *Trends Genet.*, **1**, 224.
102. Weiher,H., Konig,M. and Gruss,P. (1983) Multiple point mutations affecting the simian virus 40 enhancer. *Science*, **219**, 626.
103. Pettersson,M. and Schaffner,W. (1987) A purine-rich DNA sequence motif present in SV40 and lymphotrophic papovavirus binds a lymphoid-specific factor and contributes to enhancer activity in lymphoid cells. *Genes Dev.*, **1**, 962.
104. Wildeman,A.G., Sassone-Corsi,A.G., Grundstrom,T., Zenke,M. and Chambon,P. (1984) Stimulation of *in vitro* transcription from the SV40 early promoter by the enhancer involves a specific *trans*-acting factor. *EMBO J.*, **3**, 3129.
105. Sassone-Corsi,P., Dougherty,J., Wasylyk,B. and Chambon,P. (1984) Stimulation of *in vitro* transcription from heterologous promoters by the SV40 enhancer. *Proc. Natl. Acad. Sci. USA*, **81**, 308.
106. Sergeant,A., Bohmann,D., Zentgraf,H., Weiher,H. and Keller,W. (1984) A transcription enhancer acts *in vitro* over distances of hundreds of base pairs on both circular and linear templates but not on chromatin-reconstituted DNA. *J. Mol. Biol.*, **180**, 577.
107. Sassone-Corsi,P., Wildeman,A. and Chambon,P. (1985) A *trans*-acting factor is responsible for the simian virus 40 enhancer activity *in vitro*. *Nature*, **313**, 458.
108. Scholer,H.R. and Gruss,P. (1985) Cell type-specific transcriptional enhancement *in vitro* requires the presence of *trans* acting factors. *EMBO J.*, **4**, 3005.

109. Augereau,P. and Chambon,P. (1986) The mouse immunoglobulin heavy chain enhancer: effect on transcription *in vitro* and binding of proteins present in HeLa and lymphoid B cell extracts. *EMBO J.,* **5**, 1791.
110. Dougherty,J., Augereau,P. and Chambon,P. (1986) The mouse immunoglobulin heavy-chain gene enhancer contains sequences that inhibit transcription *in vitro* in HeLa cell extracts. *Mol. Cell. Biol.,* **6**, 4117.
111. Wasylyk,C. and Wasylyk,B. (1986) The immunoglobulin heavy chain enhancer efficiently stimulates transcription in non-lymphoid cells. *EMBO J.,* **5**, 553.
112. Imler,J.-L., Lemaire,C., Wasylyk,C. and Wasylyk,B. (1987) Negative regulation contributes to tissue specificity of the immunoglobulin heavy chain enhancer. *Mol. Cell. Biol.,* **7**, 2558.
113. Weinberger,J., Jat,P.S. and Sharp,P.A. (1988) Localization of a repressive sequence contributing to B-cell specificity in the immunoglobulin heavy chain enhancer. *Mol. Cell. Biol.,* **8**, 988.
114. Lee,W., Haslinger,A., Karin,M. and Tjian,R. (1987) Activation of transcription by two factors that bind promoter and enhancer sequences of the human metallothionein gene and SV40. *Nature,* **325**, 368.
115. Lee,W., Mitchell,P. and Tjian,R. (1987) Purified transcription factor AP-1 interacts with TPA-inducible enhancer elements. *Cell,* **49**, 741.
116. Angel,P., Imagawa,M., Chiu,R., Stein,B., Imbra,R.J., Rahmsdorf,H.J., Jonat,C., Herrlich,P. and Karin,M. (1987) Phorbol ester-inducible genes contain a common *cis* element recognized by a TPA-modulated *trans*-acting factor. *Cell,* **49**, 729.
117. Imbra,R.J. and Karin,M. (1986) Phorbol ester induces the transcriptional stimulatory activity of the SV40 enhancer. *Nature,* **323**, 558.
118. Maki,Y., Bos,T.J., Davis,C., Starbuck,M. and Vogt,P.K. (1987) Avian sarcoma virus 17 carries the *jun* oncogene. *Proc. Natl. Acad. Sci. USA,* **84**, 2848.
119. Arndt,K. and Fink,G. (1986) CGN4 protein, a positive transcription factor in yeast, binds general control promoters at all 5′ TGACTC 3′ sequences. *Proc. Natl. Acad. Sci. USA,* **83**, 8516.
120. Vogt,P.K., Bos,T.J. and Doolittle,R.F. (1987) Homology between the DNA-binding domain of the GCN4 regulatory protein of yeast and the carboxy-terminal region of a protein coded for by the oncogene *jun*. *Proc. Natl. Acad. Sci. USA,* **84**, 3316.
121. Struhl,K. (1987) The DNA-binding domains of the jun oncoprotein and the yeast GCN4 transcriptional activator protein are functionally homologous. *Cell,* **50**, 841.
122. Bohmann,D., Bos,T.J., Adman,A., Nishimura,T., Vogt,P.K. and Tjian,R. (1987) Human proto-oncogene c-*jun* encodes a DNA binding protein with structural and functional properties of transcription factor AP1. *Science,* **238**, 1386.
123. Angel,P., Allegretto,D., Okino,S., Hattor,K., Boyle,W.J., Hunter,T. and Karin,M. (1988) Oncogene *jun* encodes a sequence specific *trans*-activator similar to AP-1. *Nature,* **332**, 166.
124. Bos,T.J., Bohmann,D., Tsuchie,H., Tjian,R. and Vogt,P.K. (1988) v-*jun* encodes a nuclear protein with enhancer binding properties of AP1. *Cell,* **52**, 705.
125. Harshman,K.D., Moye-Royley,W.S. and Parker,C.S. (1988) Transcriptional activation by the SV40 AP1 recognition element in yeast is mediated by a factor similar to AP-1 that is distinct from GCN4. *Cell,* **53**, 321.
126. Jones,R.H., Moreno,S., Nurse,P. and Jones,N.C. (1988) Expression of the SV40 promoter in fission yeast: identification and characterisation of an AP-1-like factor. *Cell,* **53**, 659.
127. Veldman,G.M., Lupton,S. and Kamen,R. (1985) Polyoma virus enhancer contains multiple redundant sequence elements that activate both DNA replication and gene expression. *Mol. Cell. Biol.,* **5**, 649.
128. Piette,J. and Yaniv,M. (1987) Two different factors bind to the α-domain of the polyoma virus enhancer, one of which also interacts with the SV40 and c-*fos* enhancers. *EMBO J.,* **6**, 1331.
129. Kryske,M.H., Piette,J. and Yaniv,M. (1987) Induction of a factor that binds to the polyoma virus A enhancer on differentiation of embryonal carcinoma cells. *Nature,* **32**, 254.
130. Teich,N.M., Weiss,R.A., Martin,G.R. and Lowy,D.R. (1977) Virus infection of murine teratocarcinoma stem cell lines. *Cell,* **23**, 973.
131. Katinka,M., Vasseur,M., Montreau,N., Yaniv,M. and Blangy,P. (1981) Polyoma DNA

sequences involved in control of viral gene expression in murine embryonal carcinoma cells. *Nature*, **290**, 720.

132. Wildeman,A.G., Zenke,M.,Schatz,C., Wintzerith,M., Grundstrom,T., Matthes,H., Takahashi,K. and Chambon,P. (1986) Specific protein binding to the simian virus 40 enhancer *in vitro. Mol. Cell. Biol.*, **6**, 2098.

133. Davidson,I., Fromental,C., Augereau,P., Wildeman,A., Zenke,M. and Chambon,P. (1986) Cell-type specific protein binding to the enhancer of simian virus 40 in nuclear extracts. *Nature*, **323**, 544.

134. Mitchell,P.J., Wang,C. and Tjian,R. (1987) Positive and negative regulation of transcription *in vitro*: enhancer binding protein AP-2 is inhibited by SV40 T antigen. *Cell*, **50**, 847.

135. Sen,R. and Baltimore,D. (1987) Inducibility of $\varkappa$ immunoglobulin enhancer-binding protein NF-$\varkappa$B by a posttranslational mechanism. *Cell*, **47**, 921.

136. Baldwin,A.S.,Jr and Sharp,P.A. (1987) Binding of a nuclear factor to a regulatory sequence in the promoter of the mouse H2K^b class I major histocompatibility gene. *Mol. Cell. Biol.*, **7**, 305.

137. Imagawa,M., Chiu,R. and Karin,M. (1987) Transcription factor AP-2 mediates induction by two different signal transduction pathways: protein kinase C and cAMP. *Cell*, **51**, 251.

138. Chiu,R., Imagawa,M., Imbra,R.J., Bockoven,J.R. and Karin,M. (1987) Multiple *cis* and *trans*-acting elements mediate the transcriptional response to phorbol esters. *Nature*, **329**, 648.

139. Xiao,J.H., Davidson,I., Macchi,M., Rosales,R., Vigneron,M., Staub,A. and Chambon,P. (1987) *In vitro* binding of several cell-specific and ubiquitous nuclear proteins to the GT-I motif of the SV40 enhancer. *Genes Dev.*, **1**, 794.

140. Xiao,J.H., Davidson,I., Ferrandon,D., Rosales,R., Vigneron,M., Macchi,M., Ruffernach,F. and Chambon,P. (1987) One cell-specific and three ubiquitous nuclear proteins bind *in vitro* to overlapping motifs in the domain B1 of the SV40 enhancer. *EMBO J.*, **6**, 3005.

141. Mermod,N., Williams,T.J. and Tjian,R. (1988) Enhancer binding factors AP-4 and AP-1 act in concert to activate SV40 late transcription *in vitro. Nature*, **332**, 557.

142. Weber,F., deVilliers,J. and Schaffner,W. (1983) An SV40 'enhancer trap' incorporates exogenous enhancers or generates enhancers from its own sequences. *Cell*, **36**, 983.

143. Ephrussi,A., Church,G.M., Tonegawa,S. and Gilbert,W. (1985) B lineage-specific interactions of an immunoglobulin enhancer with cellular factors *in vivo. Science*, **227**, 134.

144. Sen,R. and Baltimore,D. (1986) Multiple nuclear factors interact with the immunoglobulin enhancer sequences. *Cell*, **46**, 705.

145. Lenardo,M., Pierce,J.W. and Baltimore,D. (1987) Protein-binding sites in Ig gene enhancers determine transcriptional activity and inducibility. *Science*, **236**, 1573.

146. Pierce,J.W., Lenardo,M. and Baltimore,D. (1988) Oligonucleotide that binds nuclear factor NF-$\varkappa$B acts as a lymphoid specific and inducible enhancer element. *Proc. Natl. Acad. Sci. USA*, **85**, 1482.

147. Atchison,M.L. and Perry,R.P. (1987) The role of the $\varkappa$ enhancer and its binding factor NF-$\varkappa$B in the developmental regulation of $\varkappa$ gene transcription. *Cell*, **48**, 121.

148. Nabel,G. and Baltimore,D. (1987) An inducible transcription factor activates expression of human immunodeficiency virus in T cells. *Nature*, **326**, 711.

149. Singh,H., Le Bowitz,J.H., Baldwin,A.S. and Sharp,P.A. (1988) Molecular cloning of an enhancer binding protein: isolation by screening of an expression library with recognition site DNA. *Cell*, **52**, 415.

150. Israel,A., Kimura,A., Kieran,M., Yano,O., Kanellopoulos,J., LeBail,O. and Kourilsky,P. (1987) A common positive *trans*-acting factor binds to enhancer sequences in the promoters of mouse H-2 and β-2 microglobulin genes. *Proc. Natl. Acad. Sci. USA*, **84**, 2653.

151. Baldwin,A.S. and Sharp,P.A. (1988) Two transcription factors, NF-$\varkappa$B and H2TF1, interact with a single regulatory sequence in the class 1 major histocompatibility complex promoter. *Proc. Natl. Acad. Sci. USA*, **85**, 723.

152. Pinkert,C.A., Ornitz,D.M., Brinster,R.L. and Palmiter,R.D. (1987) An albumin enhancer located 10 kb upstream functions along with its promoter to direct efficient, liver-specific expression in transgenic mice. *Genes Dev.*, **1**, 268.

153. Heard,J.M., Herbomel,P., Ott,M.-O., Mottura-Rollier,A., Weiss,M. and Yaniv,M. (1987) Determinants of rat albumin promoter tissue-specificity analyzed by an improved transient expression system. *Mol. Cell. Biol.*, **7**, 2425.
154. Groski,K., Carneiro,M. and Schibler,U. (1987) Tissue-specific *in vitro* transcription from the mouse albumin promoter. *Cell*, **47**, 767.
155. Lichtsteiner,S., Wuarin,J. and Schibler,J. (1987) The interplay of DNA-binding proteins on the promoter of the mouse albumin gene. *Cell*, **51**, 963.
156. Raymondjean,M., Cereghini,S. and Yaniv,M. (1988) Several distinct 'CCAAT' box binding proteins co-exist in eukaryotic cells. *Proc. Natl. Acad. Sci. USA*, **85**, 757.
157. Cereghini,S., Raymondjean,M., Carranca,A.G., Herbomel,P. and Yaniv,M. (1987) Factors involved in control of tissue-specific expression of albumin gene. *Cell*, **50**, 627.
158. Sleigh,M.J. and Lockett,T.J. (1985) SV40 enhancer activation during retinoic acid induced differentiation of F9 embryonal carcinoma cells. *EMBO J.*, **4**, 3831.
159. Cremisi,C. and Babinet,C. (1986) Negative regulation of early polyomavirus expression in mouse embryonal carcinoma cells. *J. Virol.*, **59**, 761.
160. Linney,E., Davis,B., Overhauser,J., Chao,E. and Fan,H. (1984) Non-function of a Moloney murine leukaemia virus regulatory sequence in F9 embryonal carcinoma cells. *Nature*, **308**, 470.
161. Gorman,C.M., Rigby,P.W.J. and Lane,D.P. (1985) Negative regulation of viral enhancers in undifferentiated embryonic stem cells. *Cell*, **42**, 519.
162. Hassell,J.A., Muller,W.J. and Mueller,C.R. (1986) The dual role of the polyoma enhancer in transcription and DNA replication. In *Cancer Cells, Vol. 4. DNA Tumor Viruses*. Botchan,M., Grodzicker,T. and Sharp,P.A. (eds), Cold Spring Harbor Laboratory, New York, p. 561.
163. Herbomel,P., Bourchot,B. and Yaniv,M. (1984) Two distinct enhancers with different cell specificities coexist in the regulatory region of polyoma. *Cell*, **39**, 653.
164. Ostapchuk,P., Diffley,J.F.X., Bruder,J.T., Stillman,B., Levine,A.J. and Hearing,P. (1986) Interaction of a nuclear factor with the polyoma virus enhancer region. *Proc. Natl. Acad. Sci. USA*, **83**, 8550.
165. Piette,J., Kryszke,M.H. and Yaniv,M. (1985) Specific interaction of cellular factors with the B enhancer of polyoma virus. *EMBO J.*, **4**, 2675.
166. Fujimura,F.K. (1986) Nuclear activity from F9 embryonal carcinoma cells binding specifically to the enhancers of wildtype polyoma virus and Py EC mutant DNAs. *Nucleic Acids Res.*, **7**, 2845.
167. Kovesdi,I., Satake,M., Furukawa,K., Reichel,R., Ito,Y. and Nevins,J.R. (1987) A factor discriminating between the wild-type and a mutant polyoma enhancer. *Nature*, **328**, 87.
168. Speck,N.A. and Baltimore,D. (1987) Six distinct nuclear factors interact with the 75 base pair repeat of the Moloney murine leukaemia virus enhancer. *Mol. Cell. Biol.*, **7**, 1101.
169. Berk,A.J. (1986) Adenovirus promoters and E1A activation. *Annu. Rev. Genet.*, **20**, 45.
170. Borrelli,E., Hen,R. and Chambon,R. (1984) Adenovirus-2 E1A products repress enhancer-induced stimulation of transcription. *Nature*, **312**, 608.
171. Velcich,A. and Ziff,E. (1985) Adenovirus E1a proteins repress transcription from the SV40 early promoter. *Cell*, **40**, 705.
172. Hen,R., Borrelli,E., Fromental,C., Sassone-Corsi,P. and Chambon,P. (1986) A mutated polyoma virus enhancer which is active in undifferentiated embryonal carcinoma cells is not repressed by adenovirus 2 E1A products. *Nature*, **321**, 249.
173. Sassone-Corsi,P., Fromental,C. and Chambon,P. (1987) A *trans*-acting factor represses the activity of the polyoma virus enhancer in undifferentiated embryonal carcinoma cells. *Oncogene Res.*, **1**, 113.
174. Sleigh,M.J., Lockett,T.J., Kelly,J. and Lewy,D. (1987) Competition studies with repressors and activators of viral enhancer function in F9 mouse embryonal carcinoma cells. *Nucleic Acids Res.*, **15**, 4307.
175. Imperiale,M.J., Kao,H.T., Feldman,L., Nevins,J. and Strickland,S. (1984) Common control of the heat shock gene and early adenovirus genes: evidence for a cellular E1A-like activity. *Mol. Cell. Biol.*, **4**, 867.
176. Kovesdi,I., Reichel,R. and Nevins,J.R. (1986) E1A transcription induction: enhanced binding of a factor to upstream protein sequences. *Science*, **231**, 719.
177. Kovesdi,I., Reichel,R. and Nevins,J.R. (1987) Role of an adenovirus E2 promoter

binding factor E1A mediated coordinate gene control. *Proc. Natl. Acad. Sci. USA,* **84**, 2180.

178. Reichel,R., Kovesdi,I. and Nevins,J.R. (1987) Developmental control of a promoter-specific factor that is also regulated by the E1A gene product. *Cell,* **48**, 501.

179. La Thangue,N.B. and Rigby,P.W.J. (1987) An adenovirus E1A-like transcription factor is regulated during the differentiation of murine embryonal carcinoma stem cells. *Cell,* **49**, 507.

180. Gehring,W.J. and Hiromi,Y. (1986) Homeotic genes and the homeobox. *Annu. Rev. Genet.,* **20**, 147.

181. Colberg-Poley,A.M., Voss,S.D., Chowdhury,K. and Gruss,P. (1985) Structural analysis of murine genes containing homeo box sequences and their expression in embryonal carcinoma cells. *Nature,* **314**, 713.

182. Page,D.C., Mosher,R., Simpson,E.M., Fisher,E.M.C., Mardon,G., Pollack,J., McGillivray,B., de la Chapelle,A. and Brown,L.G. (1987) The sex-determining region of the human Y chromosome encodes a finger protein. *Cell,* **51**, 1091.

183. Schuh,R., Aicher,W., Gaul,U., Cote,S., Preiss,A., Maier,D., Seifert,E., Nauber,U., Schroder,C., Kemler,R. and Jackle,H. (1986) A conserved family of nuclear proteins containing structural elements of the finger protein encoded by Kruppel, a *Drosophila* segmentation gene. *Cell,* **47**, 1025.

184. Rosenberg,U.B., Schroder,C., Preiss,A., Kienlin,A., Cote,S., Riede,I. and Jackle,H. (1986) Structural homology of the product of the *Drosophila* Kruppel gene with *Xenopus* transcription factor IIIA. *Nature,* **319**, 336.

185. Chowdhury,K., Deutsch,U. and Gruss,P. (1987) A multigene family encoding several 'finger' structures is present and differentially active in mammalian genomes. *Cell,* **48**, 771.

186. Kakidani,H. and Ptashne,M. (1988) GAL4 activates gene expression in mammalian cells. *Cell,* **52**, 161.

187. Webster,N., Jin,J.R., Green,S., Hollis,M. and Chambon,P. (1988) The yeast UASg is a transcriptional enhancer in human HeLa cells in the presence of the GAL4 trans-activator. *Cell,* **52**, 169.

188. Struhl,K. (1984) Genetic properties and chromatin structure of the yeast gal regulatory element: an enhancer-like sequence. *Proc. Natl. Acad. Sci. USA,* **81**, 7865.

189. Guarente,L. and Hoar,E. (1984) Upstream activation sites of the CYC1 gene of *Saccharomyces cerevisiae* are active when inverted but not when placed downstream of the TATA box. *Proc. Natl. Acad. Sci. USA,* **81**, 7860.

190. Giniger,E., Varnum,S.M. and Ptashne,M. (1985) Specific DNA binding of GAL4, a positive regulatory protein of yeast. *Cell,* **40**, 767.

191. Ma,J. and Ptashne,M. (1987) Deletion analysis of GAL4 defines two transcriptional activating segments. *Cell,* **48**, 847.

192. Hope,I.A. and Struhl,K. (1986) Functional dissection of a eukaryotic transcriptional activator protein, GCN4 of yeast. *Cell,* **46**, 885.

193. Brent,R. and Ptashne,M. (1985) A eukaryotic transcriptional activator bearing the DNA specificity of a prokaryotic repressor. *Cell,* **43**, 729.

194. Pinkham,J., Olesen,J. and Guarente,L. (1987) Sequence and nuclear localization of the yeast HAP2 protein, an activator of transcription. *Mol. Cell. Biol.,* **7**, 578.

195. Pfeifer,K., Arcangioli,B. and Guarente,L. (1987) Yeast HAP1 activator competes with the factor RC2 for binding to the upstream activation site UAS1 of the CYC1 gene. *Cell,* **49**, 9.

196. Pfeifer,K., Prezant,T. and Guarente,L. (1987) Yeast HAP1 activator binds to two upstream activation sites of different sequence. *Cell,* **49**, 19.

197. Olesen,J., Hahn,S. and Guarente,L. (1987) Yeast HAP2 and HAP3 activators both bind to the CYC1 upstream activation site, UAS2, in an interdependent manner. *Cell,* **51**, 953.

198. Forsburg,S.L. and Guarente,L. (1988) Mutational analysis of upstream activation sequence 2 of CYC1 gene of *Saccharomyces cerevisiae*: a HAP2- HAP3-responsive site. *Mol. Cell. Biol.,* **8**, 647.

199. Chodosh,L.A., Baldwin,A.S., Carthew,R.W. and Sharp,P.A. (1988) Human CCAAT-binding proteins have heterologous subunits. *Cell,* **53**, 11.

200. Chodosh,L.A., Olesen,J., Hahn,S., Baldwin,A.S., Guarente,L. and Sharp,P.A. (1988) A yeast and a human CCAAT-binding protein have heterologous subunits that are functionally interchangeable. *Cell,* **53**, 25.

201. Brand,A.H., Micklem,G. and Nasmyth,K. (1987) A yeast silencer contains sequences that can promote autonomous plasmid replication and transcriptional activation. *Cell,* **51**, 709.
202. Shore,D. and Nasmyth,K. (1987) Purification and cloning of a DNA binding protein from yeast that binds to both silencer and activator elements. *Cell,* **51**, 721.
203. Ma,J. and Ptashne,M. (1987) A new class of yeast transcriptional activators. *Cell,* **51**, 113.
204. Gill,G. and Ptashne,M. (1987) Mutants of GAL4 protein altered in an activation function. *Cell,* **51**, 121.
205. Giniger,E. and Ptashne,M. (1987) Transcription in yeast activated by a putative amphipathic helix linked to a DNA binding unit. *Nature,* **330**, 670.
206. Haltiner,M.M., Smale,S.T. and Tjian,R.T. (1986) Two distinct promoter elements in the human ribosomal RNA gene identified by linker scanning mutgenesis. *Mol. Cell. Biol.,* **6**, 227.
207. Jones,M.H., Learned,R.M. and Tjian,R. (1988) Analysis of clustered point mutations in the human ribosomal RNA gene promoter by transient expression *in vivo. Proc. Natl. Acad. Sci. USA,* **85**, 669.
208. Grummt,I., Kuhn,A., Bartsch,I. and Rosenbauer,H. (1986) A transcription terminator located upstream of the mouse rDNA initiation site affects rRNA synthesis. *Cell,* **47**, 901.
209. McStay,B. and Reeder,R.H. (1986) A termination site for *Xenopus* RNA polymerase I also acts as an element of an adjacent promoter. *Cell,* **47**, 913.
210. Labhart,P. and Reeder,R.H. (1987) Ribosomal precursor 3' end formation requires a conserved element upstream of the promoter. *Cell,* **50**, 51.
211. Grummt,I. (1981) Specific transcription of mouse ribosomal DNA in a cell-free system that mimics control *in vivo. Proc. Natl. Acad. Sci. USA,* **78**, 727.
212. Grummt,I., Roth,E. and Paule,M. (1982) Ribosomal RNA transcription *in vitro* is species-specific. *Nature,* **296**, 173.
213. Learned,R.M., Cordes,S. and Tjian,R.T. (1985) Purification of characterization of a transcription factor that confers promoter specificity. *Mol. Cell. Biol.,* **5**, 1358.
214. Clos,J., Buutregeit,D. and Grummt,I. (1986) A purified transcription factor (TIFI-B) binds to essential sequences of the mouse rDNA promoter. *Proc. Natl. Acad. Sci. USA,* **83**, 604.
215. Learned,R.M., Learned,T.K., Haltiner,M.M. and Tjian,R.T. (1987) Human rRNA transcription is modulated by the coordinate binding of two factors to an upstream control element. *Cell,* **45**, 847.
216. Bateman,E. and Paule,M.R. (1986) Regulation of eukaryotic ribosomal RNA transcription by RNA polymerase modification. *Cell,* **47**, 445.
217. Kownin,P., Bateman,E. and Paule,M.R. (1987) Eukaryotic RNA polymerase I promoter binding is directed by protein contacts with transcription initiation factor and is DNA sequence-independent. *Cell,* **50**, 693.
218. Sollner-Webb,B. (1988) Surprises in polymerase III transcription. *Cell,* **52**, 153.
219. Segall,J., Matsui,T. and Roeder,R.G. (1980) Multiple factors are required for the accurate transcription of purified genes by RNA polymerase III. *J. Biol. Chem.,* **255**, 11986.
220. Lassar,A.B., Martin,P.M. and Roeder,R.G. (1983) Transcription of class III genes: formation of preinitiation complexes. *Science,* **222**, 740.
221. Engelke,D.R., Ng,S.-Y., Shastry,B.S. and Roeder,R.G. (1980) Specific interaction of a purified transcription factor with an internal control region of 5S RNA genes. *Cell,* **19**, 717.
222. Pelham,H.R.B. and Brown,D.D. (1980) A specific transcription factor that can bind either the 5S RNA gene or 5S RNA. *Proc. Natl. Acad. Sci. USA,* **77**, 4170.
223. Sakonju,S. and Brown,D.D. (1982) Contact points between a positive transcription factor and the *Xenopus* 5S RNA gene. *Cell,* **31**, 395.
224. Smith,D.R., Jackson,I.J. and Brown,D.D. (1984) Domains of the positive transcription factor specific for the *Xenopus* 5S RNA gene. *Cell,* **37**, 645.
225. Ginsberg,A.M., King,B.O. and Roeder,R.G. (1984) *Xenopus* 5S gene transcription factor TFIIIA: characterization of a cDNA clone and measurement of RNA levels throughout development. *Cell,* **39**, 479.
226. Hanas,J.S., Hazuda,D.J., Bogenhagen,D.F., Wu,F.Y.-H. and Wu,C.-W. (1983) *Xenopus*

transcription factor A requires zinc for binding to the 5S RNA gene. *J. Biol. Chem.*, **258**, 14120.

227. Miller,J., McLachlan,A.D. and Klug,A. (1985) Repetitive zinc-binding domains in the protein transcription factor IIIA from *Xenopus* oocytes. *EMBO J.*, **4**, 1609.

228. Berg,J. (1986) Potential metal-binding domains in nucleic acid binding proteins. *Science*, **232**, 485.

229. Hoeffler,W.K. and Roeder,R.G. (1985) Enhancement of RNA polymerase III transcription by the E1A gene product of adenovirus. *Cell*, **41**, 955.

230. Yoshinaga,S.K., Boulanger,P. and Berk,A.J. (1987) Resolution of human transcription factor TFIIIC into two functional components. *Proc. Natl. Acad. Sci. USA*, **84**, 3585.

231. Ullu,E. and Weiner,A.M. (1985) Upstream sequences modulate the internal promoter of the human 7SL RNA gene. *Nature*, **318**, 371.

232. Krol,A., Carbon,P., Ebel,J.P. and Appel,B. (1987) *Xenopus tropicalis* U6 Sn RNA genes transcribed by RNA polymerase III contain the promoter elements used by pol II dependent U Sn RNA genes. *Nucleic Acids Res.*, **15**, 2463.

233. Kunkel,G.R., Maser,R.L., Calvet,J.P. and Pederson,T. (1986) U6 small nuclear RNA is transcribed by RNA polymerase III. *Proc. Natl. Acad. Sci. USA*, **83**, 8575.

234. Carbon,P., Murgo,S., Ebel,J.-P., Krol,A., Tebb,G. and Mattaj,I.W. (1987) A common octamer motif binding protein is involved in the transcription of U6 snRNA by RNA polymerase III. *Cell*, **51**, 71.

235. Murphy,S., Liegro,Di.C. and Melli,M. (1987) The *in vitro* transcription of the 7SK RNA gene by RNA polymerase III is dependent only on the presence of an upstream promoter. *Cell*, **51**, 81.

236. Das,G., Henning,D., Wright,D. and Reddy,R. (1988) Upstream regulatory elements are necessary and sufficient for transcription of a U6 RNA gene by RNA polymerase III. *EMBO J.*, **7**, 503.

237. Chen,W., Tabor,S. and Struhl,K. (1987) Distinguishing between mechanisms of eukaryotic transcriptional activation with bacteriophage T7 RNA polymerase. *Cell*, **50**, 1047.

238. Grosveld,F., van Assendelft,G.B., Greaves,D.R. and Kollias,G. (1987) Position-independent, high-level expression of the human β-globin gene in transgenic mice. *Cell*, **51**, 975.

239. Hames,B.D. and Glover,D.M. (eds) (1988) *Molecular Immunology.* IRL Press, Oxford.

Enhancers and other *cis*-acting regulatory sequences

Antonis K. Hatzopoulos,
Uwe Schlokat and Peter Gruss

1. Introduction

Vertebrates develop from a single cell, the fertilized oocyte, into an amazing array of different cell-types, each performing specialized functions. Except for a few known exceptions, such as the immune system, all cells probably contain the same genetic material. Distinct phenotypes appear through differential processing of the same genetic information. In other words, genes are turned on and off selectively in different tissues as needed. The main question then is: 'What mechanisms control the expression of genetic information?' It would not be unreasonable to expect that the first steps in this expression are very important. Indeed, although subsequent steps like RNA processing, mRNA stability, translation and post-translational modifications also provide opportunities to regulate gene expression, transcriptional control has turned out to be a crucial process in many cases. Therefore, analysis of transcriptional activation in molecular detail is essential as a means to better understand differentiation and development.

It is useful at this point to briefly summarize our knowledge of transcriptional regulation of gene expression in prokaryotic systems. Conserved sequences located $10-35$ bp upstream from the transcription initiation site are required for RNA polymerase binding. These constitutive elements include the TATA or Pribnow box at position -10, and the -35 box. Additional sequences are needed for positive and negative regulation. In the case of positive control, activator proteins bind to operator sites next to the constitutive elements and stimulate transcription through direct contact with RNA polymerase. For negative regulation, a repressor protein binds to sequences overlapping the constitutive elements. This interaction hinders the RNA polymerase binding and in this way inhibits transcription. Thus the nature of the transcriptional system appears to be dual. On the one hand, *cis*-acting elements, i.e. short stretches of DNA, are present in the vicinity of a gene. On the other hand, *trans*-acting factors, i.e. cellular proteins, interact with these sequences and thereby regulate expression of the adjacent gene (for a comprehensive overview of transcriptional control in prokaryotes see ref. 1).

Information about control mechanisms in higher eukaryotes became available when techniques of manipulating genes *in vitro* and subsequently introducing them into cells were developed. These techniques, collectively called 'reverse or surrogate genetics', were extensively used to dissect the control elements of many eukaryotic genes. Initially, the situation appeared very similar to prokaryotes. There is a TATA box, about 30 bp upstream from the cap site, which is necessary for accurate positioning of the initiation site of transcription. Around position −70, additional promoter sequences, like the CCAAT box and the GGGCGG motif, are required for efficient initiation (2−4). Then came some puzzling observations. First it was shown that sequences more than 100 bp upstream from the initiation site of the histone H2A gene could act in either orientation to restore efficient transcription (5). Similarly, deletion analysis showed that sequences located more than 100 bp upstream from the major 'early' cap site of Simian virus 40 (SV40) were essential for expression of the early viral genes and consequently for the virus viability (6,7; for a description of the SV40 life cycle and for the numbering system used here see ref. 8). These results were initially surprising because there was no prokaryotic precedent for such distal control signals. Very soon it became evident that these genetic elements, called enhancers, have some extraordinary properties. For example, they can also activate transcription from heterologous promoters, with exceptional positional flexibility. Enhancers can act from a distance (sometimes several thousand basepairs), independently of orientation and either from the 5' or the 3' end of a gene (9−11). These properties are the functional (and so far the sole) criteria to define enhancer elements. These findings also helped to explain the increased transformation efficiency of the herpes simplex virus thymidine kinase gene (HSV-TK) when it was linked to SV40 sequences (12).

In recent years, enhancers have been characterized in a large number of viral and cellular genes. Most of them activate transcription in a cell-specific manner which means enhancers represent one of the key elements in differential gene expression. Lately, distinguishing enhancers from traditional promoter elements has become increasingly difficult since both groups share many properties. For example, the 21 bp repeats that are part of the SV40 promoter can also work in either orientation (13,14). The amazing positional flexibility though, remains a unique property of enhancer sequences. However, because our knowledge of transcriptional regulation is incomplete, it still remains an open question whether enhancers represent a truly novel group of *cis*-acting elements. For practical purposes, we will follow the current classification. In the next sections we will attempt to summarize available information about viral and cellular enhancers, concentrating on the best studied examples. We will refer to other regulatory elements only when necessary for a full description of the expression pattern of certain genes. At the end, we will present and discuss current models about the mechanism of enhancer-mediated transcriptional stimulation.

2. Viral enhancers

2.1 The SV40 enhancer

The SV40 enhancer was the first enhancer to be discovered (9 – 11). This genetic element has been extensively studied and today it represents the best example of a prototypical enhancer. It is located in the non-transcribed region of the viral genome, between the early and late transcriptional units (ref. 8 and *Figure 1A*). The SV40 enhancer functions in a wide variety of cell types from different species. It is even active in non-mammalian systems like *Xenopus laevis* kidney cells (15) or even in the green alga *Acetabularia* (16). In some cases, the SV40 enhancer can confer cell specificity or host-range response to the promoter it activates (17 – 20).

How strong is the SV40 enhancer? Depending on the assay system, activation can reach two to three orders of magnitude when compared to the basal level of expression from enhancerless constructs (9,10,21). This activation, at least in part, is due to an increase in the number of RNA polymerase molecules that transcribe the linked gene (22,23). In the following sections, we will analyze in more detail our current knowledge about the SV40 enhancer.

2.1.1 Properties of the SV40 enhancer: activation of heterologous promoters

The SV40 enhancer can activate transcription from both the early and late region promoters even when it is placed in new positions within the viral genome (11). It can also activate correct initiation of transcription from a large number of heterologous promoters like the β-globin promoter, the conalbumin and the adenovirus major late promoters, just to name a few (9,10). Certain promoters seem to be resistant to this activation, as in the case of α-globin (24,25) and HSV-TK genes (26). The latter will respond to the SV40 enhancer though, in cell lines that provide T-antigen or possibly adenovirus E1A protein in *trans* (26). Since the nature of induction is indirect, it could occur by different mechanisms. This could then explain why other groups have observed activation of the tk promoter by the SV40 enhancer (12,20,23).

These findings suggest that the ability of an enhancer to stimulate expression of a gene may depend on interactions between promoter and enhancers, presumably via some specific *trans*-acting factors (26). This

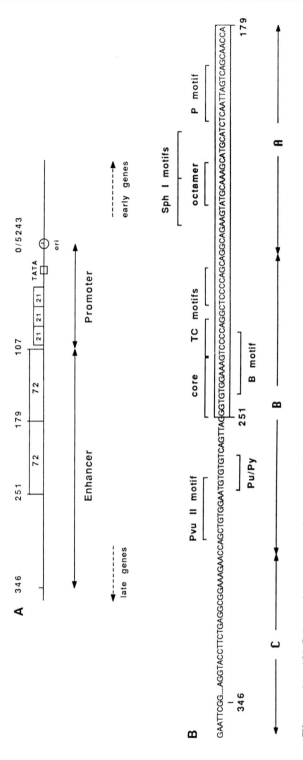

Figure 1. (A) Schematic representation of the SV40 regulatory region (numbering system according to ref. 8). The 72 bp repeats of the enhancer and the 21 bp repeats of the promoter are shown. ori represents the origin of replication, TATA denotes the location of the TATA box. Broken arrows point to the orientation of transcription of the early and late viral genes. (B) Sequence of the SV40 enhancer. Only one of the two 72 bp repeats is displayed (nucleotides 179–251, inside the rectangle). The borders of domains A, B, C are indicated at the bottom (according to ref. 21). The various motifs are shown above and below the sequence (see text). Pu/Py represents a purine-pyrimidine stretch.

concept is further supported by the fact that the immunoglobulin V_H promoter, even in the presence of the SV40 enhancer, is active in B cells but not in fibroblasts (27).

2.1.2 Distance effects

In its natural location, the SV40 enhancer is positioned immediately upstream of the promoter sequences. It can still activate transcription when placed further away, but the stimulation level decreases as the distance of the SV40 enhancer from the promoter increases (9,28). There seems to be a biphasic pattern in this drop of activity for both homologous and heterologous promoters. Insertion of small DNA fragments (up to 150 bp) leads to a dramatic decrease of enhancement. After a certain distance of about 600 bp, further insertions have only a minor effect (28). Weak promoters seem to be more prone to distance effects. An example is the promoter of the gene coding for the immunoglobulin λ light chain gene which can be activated only when the SV40 enhancer is located within 150 bp upstream from the cap site (29).

The results presented above can be explained if we consider interactions between promoter and enhancer *trans*-acting factors. When distance increases, this interaction becomes less favorable. The fact that some residual activity remains, suggests that proteins still interact, perhaps by looping out of the intervening DNA. This idea is further supported by experiments which show that inserting odd multiples of half DNA turns between the SV40 enhancer and promoter reduces activity more than inserting an even number of turns (30). In the first case, proteins bound to enhancer and promoter DNA will face opposite sites of the helix and protein–protein interactions are unfavorable.

The effect of the SV40 enhancer on a certain promoter also depends on the nature of intervening sequences. It has been shown that the enhancer preferentially potentiates transcription from the most proximal promoter (31,32). In some cases, transcription was activated from 'pseudo' promoters found in plasmid sequences (31). It is not yet clear if this is generally true, since experiments with the immunoglobulin κ chain enhancer show no preferential activation of the most proximal promoter (33; see below). Certainly, elucidation of the situation is very important since any explanation of the mechanism of enhancer function has to take into account these apparently controversial results.

2.1.3 Mutational analysis of the SV40 enhancer

The interesting properties of the SV40 enhancer made it a favorite subject for numerous studies. Extensive analysis was focused on the structural elements responsible for activity. Its main feature is the two 72 bp repeats that extend from position 107 to 251, but sequences further upstream up to position 350 are also important for optimal activity (ref. 21; see *Figure 1A*). A repeat *per se* is not required, since one copy retains about 50%

of the original activity and the virus grows effectively (10,34). Moreover, not all enhancers contain repeats (15,35–38).

Comparison of different enhancer sequences showed that they do not share any extensive homologies, but the search pinpointed some conserved short motifs. The first, called the core (GTGTGGAAAG), is located at the 5′ end of the repeats (ref. 39; see *Figure 1B*). Mutations in this area severely affect enhancer activity, especially a G to A transition of the underlined nucleotide (39). Variants of this sequence [consensus (G)TGG$^A/_T$$^A/_T$$^A/_T$(G)] are found in many enhancer elements. A closely related sequence, the 'pseudo-core' or Pvu II motif, is located just upstream of the 72 bp repeats (*Figure 1B*). The second motif, ATGCAAAG, is a close variant of the octamer sequence ATGCAAAT found in the promoter and enhancer regions of other genes (27,40–47). Interestingly, the octamer is necessary and possibly sufficient for tissue-specific expression of the immunoglobulin genes (40).

Genetic analysis provided further insight into the structure of the SV40 enhancer. Mutations that impaired activity were engineered and revertants that restored function were isolated (34,48). In summary, the results showed that the SV40 enhancer can be dissected into three functional units each of which can cooperate with the others or with itself, if duplicated, to enhance transcription. One unit contains the core, the other, the Pvu II motif, and the third the octamer motif.

Chambon and co-workers have undertaken an extensive analysis of the sequence requirements of the SV40 enhancer (21). Using systematic deletions and point mutations spanning the entire enhancer region, they reached the following conclusions.

(i)   No single nucleotide change can decrease enhancer activity by more than 6- to 8-fold, the lower activity still representing a 50-fold increase over enhancerless constructs.

(ii)   The SV40 enhancer can be divided into three broad domains called A, B and C (*Figure 1B*). Domains A and B contain a plethora of short sequence motifs, each contributing to a different extent in enhancer activity.

Domain A contains two almost perfect repeats of nine nucleotides called the Sph I motifs. The juxtaposition of these two repeats creates a close variant of the octamer motif discussed above. Domain A also contains the P motif (TCAATTAGTCA) found in polyoma enhancer A where it partially overlaps with the adenovirus E1A-like core (49,50; see below).

The core (or GTI) motif and the Pvu II (or GTII) motif are found in domain B. The core motif is followed by repetition of the sequence TCCCAG (the TC motifs) of which only the upstream one is important for enhancer activity (21). The core and the upstream TC motif together create a site which is also found in the immunoglobulin ϰ light chain gene enhancer. This site is recognized by Nuclear Factor ϰB (NF-ϰB) which is found only in lymphoid cells that express the ϰ gene (51; see below and Chapter 1).

The Pvu II motif is followed by a stretch of alternating purines/pyrimidines. It has been suggested that such sequences contribute to enhancement by forming Z-DNA structures (52), but mutations in this area have only a minor effect (21). Both domains A and B are needed for maximal activity, but if either one is duplicated, it can functionally replace the other. Domain A can also be replaced by polyoma enhancer A. Moreover, the relative orientation and position of the two domains are not crucial. These observations suggest that domains A and B are independent enhancer elements (21).

Finally, domain C extends from position 308 to 350. The importance of these sequences becomes apparent only when domains A and B are truncated (21). This observation helped explain results from previous reports about generation of functional enhancers when this region was duplicated (53,54). Domain C contains the Pu box and the KID box (55). The Pu box, a 12 bp long purine-rich sequence, is also present in the lymphotropic papovavirus enhancer and contributes to its activity in lymphoid cells. The KID box confers activity in kidney CV-1 cells.

2.1.4 Stimulatory effects of SV40 enhancer in vitro: interaction with cellular proteins

The properties of the SV40 enhancer observed *in vivo* have been reproduced to some extent *in vitro* using nuclear extracts and purified DNA templates (56 – 58). The stimulatory effect, 5- to 12-fold, is much lower than the one *in vivo*. The enhancer can still act in either orientation but not from a distance. These results suggest that *trans*-acting factors are responsible for enhancer function. Competition experiments using SV40 enhancer sequences as competitor DNA *in vivo* showed that activity could be titrated out, indicating that the presence of these factors is required for enhancement (59). The competition was conferred by enhancer but not by promoter sequences. Mutated enhancer sequences that were functionally defective also failed to compete for cellular components. Similar results were obtained with competition experiments *in vitro* (57,58,60).

The SV40 enhancer is hypersensitive to digestion by DNase I (11,61). The region in and around the enhancer does not contain nucleosomes but rather the area appears as a nucleosomal gap in electron micrographs of viral mini-chromosomes (62). The gap is an intrinsic property of the enhancer sequences, since moving the enhancer to a new location in the viral genome also induces a gap in the new site. Moreover, using two copies of the enhancer increases the length of the gap respectively. This open configuration is most likely due to the binding of regulatory proteins to the enhancer region.

The interaction of *trans*-acting factors with the enhancer sequences was demonstrated in a more direct way using techniques that can probe DNA – protein binding, mainly DNase I footprinting (63,64) and gel retardation

(65–67). In summary, all the sequences that were singled out by mutational analysis as important were also shown to bind specific factors. The binding to these sequences is similar but not identical in various cell lines which indicates that different proteins bind to the same sequence motif (64–67). A number of these proteins have been already purified. One, called Activator Protein 1 (AP-1, size ~45 kd), binds to the P-motif as well as to the basal level enhancer of the human metallothionein gene (68). It seems that AP-1 interacts with Sp1, the promoter-specific factor that binds to the 21 bp repeats (14,69–75; for a review of Sp1 and other promoter binding factors, see ref. 4 and Chapter 1). AP-1 is related to the product of oncogene v-*jun* of the avian sarcoma virus (76). The cellular oncogene c-*jun* apparently encodes AP-1 (77,78). v-*jun* also shares homology with the DNA-binding domain of the yeast transcriptional GCN4 (79). The DNA-binding domains of *jun* and GCN4 are functionally equivalent (80) and actually the *jun* oncoprotein can activate transcription in yeast (81). This finding provides direct evidence that the mechanisms of transcriptional activation have been widely preserved during evolution.

A protein interacting with the TC-motifs called AP-2 (50 kd) has also been purified (82). Interestingly, AP-2 binding is blocked by the SV40 T antigen. This inhibition is mediated by protein–protein interactions and provides a possible molecular mechanism for the transcriptional inactivation of the early viral genes by T antigen (82). Another factor, AP-3 (57 kd), binds to the core or GTI motif adjacently to AP-2, but it does not bind to the closely related Pvu II or GTII motif (82). A second core-binding protein called EBP20 (20 kd) has been purified from liver (83). The same protein also interacts with the CCAAT box.

Initially, a protein interacting with the octamer motif was partially purified and shown to enhance transcription (84). Recently, several of the octamer-binding factors have been isolated. The first, called OBP100 (~100 kd), is ubiquitous and interacts with two sites in the SV40 enhancer, the octamer ATGCAAAG and an octamer-like adjacent sequence ATGCATCT (85). The second, OTF-1 (90 kd), is also ubiquitous, while the third (three peptides 58.5, 61 and 62 kd) is lymphoid-specific (86,87). OBP100 and OTF-1 are most probably the same factor and are in turn identical to Nuclear Factor III (NFIII; see also Chapter 1).

Recently, it was shown that the stimulatory activity of the SV40 enhancer can be induced in a hepatoma cell line by the phorbol ester tumor promoter 12-*O*-tetradecanoylphorbol-13-acetate (TPA) (88). TPA also stimulates transcription by some oncogenes like c-*myc* and c-*fos* (89). Expression of these genes is increased by serum and growth factors through a mechanism involving activation of protein kinase C. Multiple SV40 *cis*-acting elements and *trans*-acting factors (AP-1, AP-2, AP-3 etc.) are implicated in this process (90). Interestingly, AP-1 is only activated by TPA while AP-2 can be additionally stimulated by cAMP through a pathway involving activation of protein kinase A (91). Further analysis of the TPA effect on the SV40 enhancer will provide an insight in areas related to cell growth and oncogenesis.

2.2 The polyoma enhancer

Polyoma is a small DNA virus that belongs to the same group of papova-viruses as SV40. Both viruses have similar size and genomic organization (8,92). As one would expect, the corresponding non-transcribed region of the polyoma genome is required for expression of the viral early genes (93). This region, located between the *Bcl*I site at position 5021 to the *Pvu*II site at position 5262, or about 180 bp upstream from the major early cap site, was shown to possess enhancer properties (94; numbering system according to refs 8 and 95).

2.2.1 Properties of the polyoma enhancer

The polyoma (Py) enhancer has clear differences from SV40. For example, the Py enhancer is slightly more active in mouse cells, while the SV40 enhancer is 4–6 times more active in primate cells (18). The two enhancers do not share extensive homology and the most well characterized wild type polyoma, strain A2, has no apparent repeats (8). Structural analysis showed that the Py enhancer can be divided into two parts called enhancers A and B (*Figure 2*; see refs 50,96). The simplest wild type polyoma contains one copy each of A and B enhancers. Both can act independently and, interestingly, they show different cell specificities. Enhancer A is 3 times stronger than B in mouse fibroblasts. Enhancer B is equally active in fibroblasts and embryonic carcinoma (EC) cells while A is 3.5 times less active in EC cells than fibroblasts (50), which might explain why enhancer A is dispensable in F9 cells (97).

Mutational analysis of the Py enhancer revealed a modular type of organization (49,98). The various elements, named α to ϵ are shown in *Figure 2*. Element α contains the core of enhancer A and is homologous to a crucial sequence repeated twice in the adenovirus E1A enhancer (99;

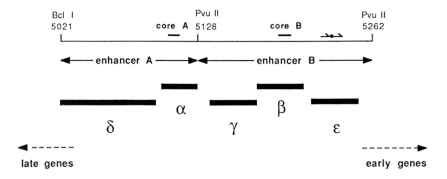

Figure 2. The polyoma enhancer (numbering system as in refs 8 and 95). The borders of enhancer A and B are shown. Solid bars indicate the location of enhancer elements α to ϵ (see text). Bars above the enhancer denote the positions of core A and core B. Half arrows above the line represent the two BPV repeats and the dot between them the location of the single base change in mutant Py ECF9.1. Broken arrows point to the orientation of transcription of early and late viral genes.

see also below). Element β contains the core of enhancer B and is homologous to the SV40 core (39). Element γ includes a GC-palindrome and a homology to the immunoglobulin enhancer. Element δ covers the late site of the enhancer and element ϵ contains the two bovine papilloma virus (BPV) repeats (100) with the F9 single base mutation between them (see below).

Two factors, PEA1 and PEA2, interact with element α (101). PEA1 is the mouse equivalent of AP-1. A single factor interacting with the Py enhancer, called PEB1, has been described (102–105). This factor binds to the GC-palindrome, but surrounding sequences, the SV40 core and the immunoglobulin enhancer homology are required for efficient binding as well. This factor also interacts with the lymphotropic papovavirus (LPV) enhancer but it cannot be competed out by polyoma enhancer A, murine sarcoma virus (MSV) or SV40 enhancers (106).

The Py enhancer is essential for viral replication as well. Interestingly, the sequences necessary for replication are the same as the ones required for transcription (49,93,96,107–109). Other enhancers (e.g. SV40) can substitute for the Py enhancer in replicatory function (107,110). Moreover, the immunoglobulin enhancer confers tissue-specific replication (107). It is possible that enhancement and replication share, at least partially, a common pathway. This hypothesis is further supported by the fact that an isolated protein which binds to the CCAAT box turned out to be Nuclear Factor I (NFI) which is involved in adenovirus replication (111). Moreover, one of the octamer-binding proteins is NFIII, which is also involved in adenovirus replication (86).

Recently, recombinant viral genomes carrying various enhancers in the place of enhancer B have been constructed (112). These recombinants show a wider tissue specificity of replication and expression of viral genes than wild type polyoma. For example, replacing enhancer B with the Moloney murine leukemia virus (MoMLV) or the immunoglobulin enhancers allows replication of polyoma in lymphoid cells. Sometimes the opposite effect is observed. For example, replacing B with the SV40 enhancer abolishes replication in all cell lines tested (112). Surprisingly, some combinations create enhancers with novel properties, that is a tissue specificity different from the parental enhancers. For example, Py A and MoMLV create a pancreas-specific enhancer (113). These findings further support a modular mechanism for the tissue specificity of enhancer function as will be discussed in Sections 4 and 5.

2.2.2 Polyoma enhancer activity in mouse embryonal teratocarcinoma cells

The Py enhancer has been the subject of numerous studies mainly because of the strict host-range specificity of this virus (114). Wild type polyoma does not productively infect murine embryonal carcinoma cells (114). This inability correlates with the inactivity of the Py enhancer in these cells (50,115). Upon differentiation, EC cells become permissive to polyoma infection and the enhancer is also active (116–118). Polyoma mutants that infect undifferentiated EC cells have been isolated (119–123). All mutants carry alterations within the Py enhancer. Similarly, mutants have been isolated that infect murine neuroblastoma cells (124), Friend erythroleukemia cells (125,126) and murine trophoblast cells (127). These mutants also carry sequence modifications within the enhancer. The alterations are characteristic and specific for the cell line in which the virus was adapted to grow. For example, polyoma mutants that grow in F9 cells are characterized by point mutations and tandem duplications of the region containing the alteration (119–123). The simplest change in the Py enhancer (Py ECF9.1) is an A to G transition at nucleotide 5233 (121–123). This base is located between the BPV repeats (*Figure 2*). PCC4 mutants usually carry deletions of enhancer B and duplications and/or translocations of enhancer A (120,128). Polyoma strains that grow in trophoblast cells have deletions in the region between A and B (127). Mutants that grow in Friend erythroleukemia and neuroblastoma cells contain duplications within the A segment (124–126). These observations make polyoma an interesting system to study, because elucidation of the molecular mechanisms that allow mutants to grow in non-permissive cells will shed light on corresponding cellular events that take place during mouse embryogenesis and cell differentiation. Besides polyoma, other viral enhancers are not active in EC cells (129). This led to a suggestion that negative factors are present in these cells, which are then lost during differentiation. This hypothesis is supported by the fact that an adenovirus E1A-like protein is present in EC cells (130). It is known that E1A protein can repress activity of some viral (131,132) and cellular enhancers (133). Interestingly, E1A does not suppress Py ECF9.1 in undifferentiated F9 cells (134). So far though, interaction of negative factors with polyoma enhancer has not been directly demonstrated. Alternatively, EC cells might be missing some positive transcriptional factors. A *trans*-acting factor has been identified in undifferentiated F9 cells which binds to the Py ECF9.1 mutant but not to the wild-type sequence (135). Apparently the point mutation has created a binding site for a positive transcription factor already present in F9 cells. In addition, it was shown that c-Ha-*ras* and TPA stimulate the Py enhancer in F9 cells (136). This stimulation does not involve reversal of an E1A-like inhibition but rather an activation of the PEA1 transcriptional factor (137).

2.3 Enhancers in other papovaviruses

2.3.1 The human BK virus enhancer

The BK virus shares about 80% sequence homology with SV40 (138,139). Sequence differences are concentrated in the non-transcribed regions. BK also harbors an enhancer element in this area, but it is not as strong as the SV40 enhancer (140). The BK enhancer consists of a 68 bp triplication. Each repeat contains a copy of the SV40 core sequence and a CCTCCC motif, similar to the Sp1 binding site. Interestingly, the BK enhancer is stimulated by E1A protein (141), while the SV40 enhancer activity is suppressed (131,132). An additional difference between the two enhancers is that only the BK enhancer interacts with NFI (142).

2.3.2 The lymphotropic papovavirus enhancer

The recently isolated lymphotropic papovavirus (143) infects only cells of primate B-lymphocyte origin (144). Sequence analysis showed that the LPV genome is very similar to the other polyoma viruses mentioned above (145). LPV harbors an enhancer, but in contrast to the enhancers discussed so far, the LPV regulatory element is tissue-specific, being active only in cells of the hematopoietic system and not in cells of fibroblast or epithelial origin (146). The main part of the LPV enhancer is two 63 bp direct repeats each containing a copy of the SV40 core. Next to the core there is a copy of the Pu box which binds a lymphoid-specific factor and is at least partially responsible for the tissue-specificity of the LPV enhancer (55).

2.3.3 The JC virus enhancer

The human papovavirus JC is very similar to SV40 and BK viruses but it has a very restricted host-tissue range since it grows only in human fetal glial cells (147). JCV is responsible for the demyelinating fatal disease progressive multifocal leukoencephalopathy (147). The JCV enhancer consists of two 98 bp repeats. Its activity is 20 times higher in fetal glial cells than HeLa cells and correlates well with the virus specificity (148). The 98 bp repeats share significant homology with a 82 bp long brain-specific identifier RNA sequence which has been found only in the introns of precursor RNA molecules isolated from rat brain (149). Although the importance of this observation is not clear, it might be related to the tissue specificity of the virus. A JCV mutant has been isolated that can grow in human embryonic kidney cells (150). The mutant encompasses complex rearrangements in the enhancer region which create sequences resembling the cores of SV40 and adenovirus E1A enhancers (151). Recently, several proteins present in nuclear extracts from brain were shown to interact with the JCV enhancer (152).

2.4 The bovine papilloma virus enhancer

The BPV enhancer has been localized to a 59 bp fragment which, contrary to the examples discussed above, is located at the 3' of the promoter after the polyadenylation site (153,154). The essential sequences contain two functional domains 25 bp apart (100). Both domains share sequence homology with the SV40 core. These two sequences are the BPV repeats in the polyoma enhancer. Recently, a second enhancer element has been characterized within the non-transcribed upstream regulatory region (URR) of BPV, which stimulates transcription only in the presence of the viral E2 gene product (155,156). This *trans*-activating protein binds to several sites within the URR through its C-terminal domain (157,158).

2.5 Enhancer elements in adenovirus

The adenovirus genome is fairly large; a linear double-stranded molecule of 35 000 bp (for comparison, papovaviruses are ~5000 bp and BPV is ~8000 bp long; for a review on adenovirus biology see ref. 92). An enhancer element has been localized at the extreme left end, just upstream of the gene coding for the E1A protein. The exact boundaries of the enhancer have not been accurately defined since three different groups have stressed the importance of different upstream sequences (99, 159 – 162). Taken together, these sequences are located from 45 to 350 bp upstream of the adenovirus E1A cap site or at positions 150 to 455 of the viral genome (numbering system as in ref. 163). The discrepancy might be due to the fact that enhancers have a modular structure and, as a result, the importance of certain motifs might be underestimated in different assay systems.

Further analysis showed that the E1A enhancer contains two domains (164). One, called enhancer I, is repeated twice at positions 200 and 300 of the viral genome and is specific for regulation of E1A gene transcription. This sequence was mentioned above as the adenovirus-like core of the polyoma enhancer (Section 2.2.1) and is duplicated in a number of wild type polyoma strains (165). In addition, it is found in some retroviral long terminal repeats (LTR) and in the regulatory region of interferon genes (99,166). Although this element is required for polyoma replication (49), it is dispensable for adenovirus replication (164). The second domain, called enhancer II, is found between the repeats of enhancer I and is required for regulation of transcription of all the early viral genes (164). Recently, a cellular protein has been identified that interacts with an SV40 core sequence within the adenovirus enhancer (167).

2.5.1 E1A regulates expression of other genes

E1A protein has many interesting properties. It is responsible for activating expression of the other adenovirus genes (168,169) and it negatively regulates its own expression (170). It can also activate cellular genes like β-globin, β-tubulin and heat shock genes (130,171 – 175), while it represses the activity of others, like the major histocompatibility class I genes (176). Interestingly, the E1A protein relieves the enhancer requirement for expression of transfected β-globin genes (171). As we discussed above, E1A affects in different ways the activity of certain enhancers as well (131 – 133,177). These effects are rather indirect and most probably E1A regulates the activity of cellular transcriptional factors (178).

One way to elucidate the mechanism of these phenomena is to study the role of E1A in the induction of the other adenovirus genes like E2.

It has been shown that promoter sequences of this gene are required for expression and stimulation by the E1A protein (179 – 181). Moreover, this promoter can act as an E1A inducible enhancer (179,182). A factor, present in HeLa cells, binds to this area only after stimulation by E1A (183 – 185). This factor is also present in F9 cells but its concentration decreases after differentiation (186,187). If E1A is introduced into differentiated F9 cells, the level of this factor increases again (186). The regulatory mechanism is probably more complex since there are reports that E1A also relieves negative regulation of the E2A promoter (188). Certainly more work is needed to clarify some controversial results but the outcome is very important for understanding transcriptional control during viral infection and the differentiation of embryonal cells.

2.6 Enhancers in herpes viruses

Enhancers have been characterized in herpes simplex virus (HSV) (189,190), herpesvirus saimiri (191), human and mouse cytomegaloviruses (192,193), and the Epstein – Barr virus (194). These viruses have large DNA genomes (150 – 225 kb) and enhancer elements are located upstream from the transcription initiation site of the immediately early genes.

The human cytomegalovirus (CMV) enhancer is among the longest and strongest enhancers identified so far, being several fold more active than SV40. It is ~400 bp long and harbors a variety of repeated sequence motifs. In particular, there are four groups of imperfect repeats, 17, 18, 19 and 21 bp long. The 17 and 21 bp repeats are found three times, the 18 bp repeat four times, and the palindromic 19 bp repeat five times (192). The 18 bp repeats share homology with the SV40 core and the 21 bp repeats contain the GGGCGG motif. As mentioned above, this motif is present in the promoter region of SV40 (13) and the HSV-TK gene (195). Mutational analysis showed that the enhancer remains active even if large deletions are performed, a result that reflects the redundancy of its sequence motifs (192). Within the CMV enhancer are several constitutive hypersensitive sites suggesting the binding of *trans*-acting factors (196). A number of factors interact with the 17, 18, 19 and 21 bp repeats (197).

The mouse CMV enhancer is very long (>700 bp) and contains a complex pattern of sequence repeats, the longest being 181 bp long (193). The 19 bp sequence is found once and the 18 bp repeat in several copies. Also present are an adenovirus enhancer homology, several CCAAT boxes and a GGGCGG motif, indicating that the CMV enhancers are complex arrays of both enhancer- and promoter-specific sequence motifs. Finally, the Epstein – Barr virus enhancer consists of 20 copies of a 30 bp direct repeat (194).

2.7 The human hepatitis B virus enhancer

Hepatitis B virus (HBV) harbors a 200 bp long enhancer in its small (3.2 kb) circular genome (198,199). The enhancer is stronger in hepatoma cells and it requires *trans*-acting cellular proteins for function (199,200). It is located about 1150 bp downstream of the major initiation site for the surface antigen and about 500 bp upstream of the core antigen gene. This part of HBV codes for the viral reverse transcriptase which makes the HBV enhancer the only enhancer found in a protein-coding region. The enhancer stimulates transcription from both the core antigen and surface antigen promoters (201).

2.8 Enhancers in retroviral long terminal repeats

The best studied example is the enhancer found in the mouse mammary tumor virus (MTV) long terminal repeats (202,203). The enhancer functions only when it is induced by glucocorticoids (204). The hormone binds to its receptor and the complex interacts with specific sites within the enhancer called glucocorticoid response elements (GRE; 202,205,206). GRE have been found in a number of glucocorticoid regulated genes (see ref. 207 and references therein). Recent evidence suggests that the hormone interaction probably exposes a hidden pre-existing site in the receptor which is responsible for binding to GRE (208). The MTV enhancer is an ideal system to study enhancement because the *cis*-acting elements have been well characterized. In addition the *trans*-acting factor (i.e. the glucocorticoid receptor) has been purified, cDNA clones and mutants are available and at this point it seems that the receptor is the only factor required for enhancement (209).

The first enhancer element in retroviral LTR was discovered in the Moloney murine sarcoma virus (MSV) (210,211). The MSV enhancer is also activated by glucocorticoids but at the same time is constitutively active (212,213). This correlates with the fact that the 73-72 repeats of MSV contain other motifs besides GRE (212,214). Binding of *trans*-acting factors has been studied more thoroughly in the closely related MoMLV (215). Six nuclear factors were identified interacting with the 75 bp repeats. One binds to an SV40-like core sequence, one to an NFI motif, one to the GRE and the other three are proteins that have not been described before. Interestingly, NFI, and to a lesser extent the core-binding protein, are depleted in embryonal carcinoma cells (215), where MoMLV regulatory sequences are not active (216). An enhancer has been also identified in Friend murine leukemia virus and it is active in erythroid cells but not in T cells where MoMLV is active (217).

The human T lymphotropic viruses HTLV-I and HTLV-II are responsible for T-cell leukemias, while HTLV-III (or LAV or HIV) is the etiological agent of the acquired immune deficiency syndrome (AIDS; see ref. 218 for references). These viruses have certain unique characteristics. For example, they possess long LTRs and an extra sequence, called the pX region, between the 3' end of the *env* gene and the LTR. In addition, the viruses are subject to *trans*-activation of the LTR driven transcription by the protein products of the pX region (see ref. 218 and references therein). Enhancers have been found in the LTR of HTLV-I (219 – 223), HTLV-II (220,224) and HTLV-III (225,226). The HTLV-I enhancer functions well in uninfected lymphoid and non-lymphoid cells and its activity seems to be independent of the transactivator protein (223,224). Other reports, though, indicate that the enhancer is required, at least partially, for the *trans*-activation of transcription (221 – 223). The HTLV-III enhancer is also active in many cell types (225). The HTLV-II enhancer, as well as the enhancer of the closely related bovine leukemia virus, work only in infected lymphoid cells (224). In a recent report, it was shown that an increase in HTLV-III gene expression correlates with the induction of NF-xB (227). As mentioned before, this cellular protein is a *trans*-acting factor interacting with the x chain enhancer and is usually found only in cells producing immunoglobulin x light chain (see also below). NF-xB has binding sites within the enhancer and it stimulates expression of the viral genes in the presence of the transactivator protein (227). Further experiments are needed to understand the mechanism of this *trans*-activation which in turn might shed light on the pathogenesis of AIDS.

Enhancers have been discovered in other retroviruses like Abelson murine leukemia virus (228), Harvey murine sarcoma virus (229), Rous sarcoma virus (230,231), and Rous associated viruses (232). In the latter case, the enhancer activity correlates with oncogenic potential. In addition, it is known that the tropism of retroviruses depends on the LTR sequences where the enhancers reside (17,217,233,234).

3.   Cellular enhancers

3.1   Enhancers in immunoglobulin genes

Identification of multiple viral enhancers subsequently led to the discovery of corresponding cellular counterparts. The first examples of such cellular enhancers were those found in the mouse immunoglobulin heavy (235 – 238), x light (239,240) and—in contrast to earlier reports (240)— supposedly λ light chain also (241). Surprisingly, at least heavy and x chain enhancers are located within the large introns between the variable and constant regions of these genes.

In order to be expressed properly, one of numerous immunoglobulin variable regions, together with its adjacent promoter, has to be placed in close proximity to the enhancer by somatic recombination during B-cell differentiation (as reviewed in ref. 242). Due to the recombinational activation of heavy and, subsequently, light chain loci, the corresponding promoters are put under the influence of the enhancers and thereby activated to give functional transcripts. The transcriptional activation in differentiating B cells correlates with the appearance of nuclease-sensitive sites in the vicinity of the enhancers, as has been shown for heavy (243) and x chain (244) loci.

3.1.1 Tissue specificity of the immunoglobulin enhancers

Following identification, careful analyses led to the localization of the heavy chain enhancer within a 1 kb *Xba*I restriction fragment which contains the full activating potential (ref. 235; see *Figure 3*). Intriguingly, in addition to distance and orientation independence, the immunoglobulin (Ig) enhancers exhibit striking cell-type specificity (146,235,236, 239 – 241); they are able to mediate transcriptional activation preferentially in cells of lymphoid origin and, in particular, in cells of B lineage (146,245). However, the enhancer elements are not the only determinants for tissue-specific expression of the immunoglobulin genes. The immunoglobulin promoters, in combination with generally active enhancers such as those from polyoma virus or SV40, also mediate cell-specific transcription (27,246 – 248), as does another intragenic region yet to be investigated more thoroughly (248). However, for maximal expression of the immunoglobulin genes, the respective immunoglobulin promoters have to act in concert with their homologous enhancers (249). These experiments have mainly involved introducing cloned genes and their variants into cells by transfection. The tissue-specific expression of Ig genes has, however, also been shown by the fusion of lymphoid cells, productively expressing immunoglobulin genes, with hepatoma cells, upon which Ig expression is rapidly lost (250).

Another approach to demonstrate the cell-specific effect of the Ig enhancer has been exploited by Brinster and associates using transgenic mice. They describe the induction of tumors restricted to lymphoid tissue by introduction of a c-*myc* oncogene linked to the IgH enhancer (251). Recently, using constructs containing conalbumin and the IgH promoter, the IgH enhancer-mediated, cell-specific regulatory effect in transgenic mice has been confirmed (252).

3.1.2 Mutational analysis of the IgH enhancer

Initial deletion mutagenesis of the IgH enhancer indirectly mapped the major enhancing activity between sites *Pst*I and *Eco*RI within the 1 kb *Xba*I intron fragment (235). Mutational fine mapping has recently been

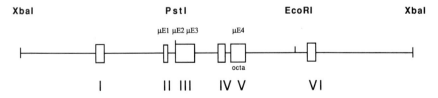

Figure 3. Schematic representation of the immunoglobulin heavy chain enhancer. Boxes I and VI represent areas of DNA – protein interactions. μE1 to μE4 indicate binding sites occupied only in B cells as mapped by genomic sequencing (according to ref. 264). *Xba*I, *Pst*I, *Eco*RI: restriction sites as referred to in the text. octa, the octamer motif as referred to in the text.

carried out. Three major conclusions were drawn.

(i)   Transcriptional potentiation dropped gradually upon deleting increasingly large portions of the enhancer (253,254). This indicates that the enhancer consists of multiple positive regulatory sequences whose additive effect results in the enhancer's overall stimulating ability. This conclusion was further supported by engineering point mutations in the μE1, E2, E3, E4 and octamer motifs (255). All mutations reduced enhancer activity to various extents, demonstrating that, as in the case with the SV40 enhancer, multiple sequence motifs contribute to optimal transcriptional activity.

(ii) The central PstI – EcoRI fragment turned out to be active in both fibroblasts and lymphoid cells, indicating deletion of negatively interfering sequences present in the undissected XbaI enhancer fragment which is unable to mediate transcriptional activation in fibroblasts. The negatively interfering sequences were mapped near to and on the 5′ side of the PstI site (possibly box II in *Figure 3*) and within the 3′ EcoRI – XbaI fragment (253,254). These flanking sequences can suppress the IgH as well as the SV40 enhancer in fibroblasts (256).

(iii) The presence of the whole PstI – EcoRI fragment, as compared to the 5′ half of it, results in a further tenfold transcriptional increase only in lymphoid cells. Therefore, the 3′ half of this fragment seems to harbor positive regulatory sequences recognized as such in B cells only. The octamer motif is present here (octa in *Figure 3*). Furthermore, the octamer by itself is sufficient for lymphoid-specific activity *in vivo* (257,258).

3.1.3 Reproduction of Ig enhancer-mediated effects in vivo by transcriptional systems in vitro

It is conceivable that enhancer sequences are activated or repressed upon binding of transcription factors, as has been demonstrated for several transcriptionally important promoter elements. Initial co-transfection competition experiments suggested the involvement of *trans*-acting soluble factors for IgH enhancer-mediated transcription (259). Additional support came from the observation that an IgH gene experimentally introduced into non-permissive cells (i.e. mouse L cells) could be induced upon microinjection of nuclear proteins prepared from B cells (260). Due to the development of nuclear protein extracts from permissive and non-permissive cells, the cell-restricted transcriptional activity of the IgH enhancer could be mimicked qualitatively, though the dramatic quantitative effect could not be observed (261,262). However, chemicals such as spermidine or polyethylene glycol seem to drastically influence the effects exerted by the Ig enhancers *in vitro* (261,263).

The large XbaI – EcoRI IgH enhancer subfragment was shown to interact with activator protein(s) also shared by SV40 and LPV (lymphotropic papovavirus) enhancers in lymphoid cells (261). *In vitro* competition experiments between SV40 and IgH enhancers in non-lymphoid cells further suggest that these cells contain positive factor(s) that can interact with both of these enhancers (60). Since the IgH enhancer (*Xba*I fragment) is silent in non-lymphoid cells *in vitro* and *in vivo* despite its ability to bind positive regulatory factors, one might postulate the existence of additional negative factors that repress its activity in non-lymphoid cells. Binding of a repressor molecule to the small *Eco*RI – *Xba*I subfragment has been suggested using a T-cell line that is permissive for IgH-mediated transcription *in vitro*. Using this fragment as a competitor led to an additional transcriptional increase probably due to titrating out the repressing molecules (261). Binding of putative repressor molecules in both lymphoid and non-lymphoid cells to the 3' end of the central *Pst*I – *Eco*RI fragment has also been proposed, based on *in vitro* transcription experiments using different deletion mutants of the latter (262).

In summary, from *in vivo* and *in vitro* transcription data, no simple model for the regulation of IgH enhancer-mediated transcription seems evident. Rather, a complex pattern of regulation emerges consisting of multiple positive factors (mainly supported by different lines of *in vivo* experiments) and negative factors (based on *in vitro* transcription experiments). It might very well be possible that it is the precise balance of positively and negatively acting transcription factors which finally results in either an overall activation or repression of the respective immunoglobulin genes in any given cell type. It remains to be determined whether any of these factors act differently within the cellular environment of different cells. It has been shown that the adenovirus type 2 E1A gene products have pleiotropic effects on the IgH enhancer. In particular, E1A leads to activation of the IgH enhancer in fibroblasts (177) while it represses the IgH enhancer activity in plasmacytomas (133). The notion that E1A-like cellular counterparts might also exert some of these effects is attractive and, indeed, the involvement of E1A-like cellular proteins in gene regulation has been postulated in embryonal carcinoma stem cells (130).

3.1.4 Interaction of Ig enhancers with trans-acting factors

Direct proof of the interaction of cellular factors with the IgH enhancer in its native environment was gained from a new technique called 'genomic sequencing' which visualizes factors bound to their respective binding sites *in vivo*. Using this approach on living cells (264) or isolated nuclei (265), the existence of multiple factors interacting with the IgH enhancer in cells of B lineage was demonstrated, whereas in cells of non-lymphoid origin no such interaction was detectable. Corresponding factors could also be identified in lymphoid nuclear protein extracts (51,262,266 – 270). Most surprisingly, and in contrast to the *in vivo* results, these factors seem to be present in cells of both lymphoid and non-lymphoid origin. This obvious discrepancy might be resolved in light of the different experimental approaches: in non-permissive cells, the IgH enhancer might exist in a tightly packed form inaccessible to the respective transcription factors, whereas in lymphoid cells these factors can bind upon local unfolding of this area. In nuclear extracts, however, the factors can easily interact with added 'naked' DNA.

As seen in *Figure 3*, six major areas of DNA–protein interaction scattered along the IgH enhancer have been identified, termed here I to VI. Owing to the different names given to seemingly identical factors in different publications, we will just refer to the unifying scheme in *Figure 3*. Binding sites μE1 to μE4, which were identified *in vivo* (264), are also shown in *Figure 3*.

A factor binding to site I has been identified in cells of lymphoid and non-lymphoid origin and binds to SV40 and Ig$\varkappa$ enhancers as well as an additional site in the IgH enhancer (271).

The factor that binds to site II is distributed uniformly in a wide variety of lymphoid and non-lymphoid cell lines from different species (268). Initial sequence comparisons of the four binding areas identified *in vivo* revealed the consensus sequence CAGGTGGC and, therefore, suggested binding of a unique factor to all sites (264). Whether factor II also interacts with the homologous μE3 binding site, however, is somewhat controversial (268,271). Interestingly, upon deletion of this binding site, the enhancer's ability to stimulate transcription in B cells decreases (254). In contrast, deletion of the same site leads to a transcriptional increase in HeLa cells (253,254). In contrast to factor I, factor II is unable to interact with the SV40 or Ig$\varkappa$ enhancers (271).

Binding area III represents a rather complex pattern of protein–DNA interaction. At least three different factors which probably bind adjacent to each other seem to be involved, one of which can interact with the SV40 Ig$\varkappa$ enhancers, another with the SV40 enhancer only, and a third with neither of the above (271). Interestingly, some of these factors also bind to MSV and polyoma enhancers which are known to exhibit a rather wide spectrum of activity (266). Furthermore, the factor that interacts with the μE3 motif does not bind to μE1 in spite of the homology between the binding sites (51).

The factor that binds to site IV does not interact with MSV or SV40 enhancers. However, it binds to a homologous sequence motif of the LPV enhancer (266). Intriguingly, this latter enhancer has been demonstrated to play a major role in the remarkable tissue tropism of this virus (146). Since the IgH enhancer functions in a subset of these cell lines, factor IV might be a good candidate for determination of the cell specificity of these two enhancers.

Site V harbors the octamer sequence also found in all immunoglobulin promoters (40,41) and many other enhancer and promoter elements

(27,42 – 47). Most interestingly, this octamer binds several distinct factors at its unique site in the IgH enhancer (266,269). One of these factors has been shown to be ubiquitously present in lymphoid and non-lymphoid cells (51,84,266 – 269), whereas another seems to be inducible in lymphoid cells upon lipopolysaccharide stimulation (51). An analogous situation has been found for the IgH promoter octamer where ubiquitous as well as mitogen-inducible factors unique to lymphoid cells (i.e. B and T cells) have been described (270). Since competition experiments have revealed that both the ubiquitous and the lymphoid-specific IgH-enhancer binding factors can also bind to the x promoter, it might very well be that an identical set of transcription factors is used in both cases (269). In contrast, another report indicates that the x promoter octamer sequence binds a slightly different set of factors (262). It is possible that sequences adjacent to the IgH enhancer and Igx promoter octamer motifs modulate the octamer's binding capability (270,272,273). Recently, the ubiquitous octamer-binding factor has been shown to be essential for SV40 enhancer (84) and human histone H2b promoter (274) function *in vitro*. Furthermore, this factor seems to stimulate adenovirus replication *in vitro* (275). It also binds to the U2 small nuclear RNA promoter (84,274) and possibly to homologous sequence motifs in the human β interferon and HSV immediate early gene promoters (see 275) as well as to the chicken lysozyme enhancer. (For a description of the purified octamer-binding factors see also Section 2.1.4 and Chapter 1.)

Finally, the factor interacting with site VI is ubiquitous and interacts also with sequences of the SV40 and Igx enhancer (271). Earlier deletion experiments confirm the contribution of this area to optimal IgH enhancer activity (236). Besides the octamer-binding factors, several other proteins interacting with the IgH enhancer have been partially purified and characterized (276).

3.1.5 Immunoglobulin x-light chain enhancer

Binding of cellular factors to the Igx enhancer has also been demonstrated, one of which seems to be ubiquitous and another specific to lymphoid cells that actively transcribe the x chain gene (51,272,277). The ubiquitous factor is also able to interact with the μE3 site in the IgH enhancer (51). The lymphoid-specific factor 'NF-xB' can be induced by the mitogen lipopolysaccharide (LPS) and the protein synthesis inhibitor cycloheximide in pre-B cells (277), an observation well correlated with the induction of x gene expression (278). This factor can also be induced in non-B cells upon phorbol ester treatment (277). These results strongly suggest that a ubiquitous precursor protein in pre-B cells can be post-translationally modified (e.g. by protein kinase C-mediated phosphorylation) into an active transcription factor. In case of the IgH enhancer, cycloheximide did not exert any significant transcriptional influence (278). Furthermore, the cellular transcription rate of the heavy chain locus has been shown to be comparably high in both pre-B and plasma cells, although there seems to be a drastic difference in the level of heavy chain transcripts present in these two cell stages (279). These observations suggest a post-transcriptional regulation (e.g. stability of messenger RNA) for immuno-globulin heavy chain production.

The presence of NF-$\varkappa$B seems to be necessary only for the establishment of enhancer-mediated transcription, whereas it might be dispensable for the maintenance of $\varkappa$ chain transcription at later stages of B-cell development (280). This is similar to a mechanism that has been suggested for SV40 enhancer action (281) where the transcriptionally active state is not abolished by DNA replication (281). This in turn might explain why some B cells can highly express their endogenous IgH chains in spite of a deletion of the corresponding enhancer element (282).

The NF-$\varkappa$B binding site acts as a lymphoid-specific and inducible enhancer, thus strengthening the notion that NF-$\varkappa$B is crucial for Ig$\varkappa$ gene expression (283). Moreover, mutational analysis showed that the rest of the sequence motifs present in the Ig$\varkappa$ enhancer also contribute to optimal activity (255).

3.2 Enhancers in genes expressed in pancreas

Pancreas consists of distinct cell types which produce different sets of proteins, e.g. insulin is synthesized in endocrine β cells while chymo-trypsin B, amylase, elastase and trypsin are produced in exocrine pancreas cells. The corresponding 5'-flanking regions of these genes are responsible for this cell-specific expression (284,285). For example, sequences at the 5' end of chymotrypsin B, amylase, elastase and trypsin genes drive expression in exocrine cells while sequences at the 5' end of the insulin gene control expression in endocrine cells (284 – 287). Moreover, chimeric constructs carrying the 5' insulin gene sequences and the SV40 oncogenes specifically induce pancreatic β-cell tumors in transgenic mice (288).

Further analysis of these 5' regions showed that enhancer elements are partly responsible for the cell-specific pattern of expression. Tissue-specific enhancers have been found at similar positions in the insulin, chymotrypsin, amylase and elastase genes (286,289,290). The 5' area of the insulin gene can be dissected into two parts. The region between $+1$ and -113 behaves as a classical promoter, while the region -103 to -353 has all the properties of an enhancer (289). Both elements work in a tissue-specific manner but the enhancer seems to play the dominant role in the tissue-specific expression of the insulin gene (289). In addition, the gene is under negative control in cells that do not normally produce insulin (291). The repression site partially overlaps both the enhancer and promoter sequences. Besides this type of negative control, it has been shown that a region located $2-4$ kb upstream from the insulin gene represses transcription (292). This sequence, termed 'silencer', belongs to the rat repetitive sequence LINES and can suppress transcription in a manner reminiscent of enhancers, i.e. it can work upstream, downstream and over heterologous promoters. These results indicate that multiple levels of control probably exist in order to ensure tight regulation of a certain gene.

Recently, attention has been focused on cellular factors interacting with the insulin enhancer. Three sites of DNA – protein contacts have been identified within the enhancer (293). The 5' most proximal site is protected only in insulin producing cells and not in heterologous cells like BHK or HeLa. The protected area contains a core sequence and is rather large (146 bp), indicating the binding of several factors (293).

3.3 The metallothionein gene enhancer

The human metallothionein gene h$MTII_A$ is transcribed in many tissues and, in addition, its expression is stimulated by glucocorticoids, heavy metals and growth factors (294 and references therein). The cis-acting sequences responsible for this regulation are located at the 5' end of the gene (295). In particular, five different elements have been characterized so far. These include a TATA box, a GGGCGG (GC) motif, two basal level enhancers (BLE), and glucocorticoid (GRE) and metal response elements (MRE).

BLE is a classical enhancer element which is probably responsible for the constitutive expression of the h$MTII_A$ gene in different tissues (296). GRE have been shown to be inducible enhancers for a number of glucocorticoid regulated genes (see Section 2.8). MREs also act as enhancers when induced by heavy metals (297).

Recently, a number of proteins have been described that interact with these regulatory sequences. These include Sp1 which binds to the GC motif and AP-1 which interacts with sequences within the BLE (68). If the GC site is mutated so that Sp1 binding is abolished, then BLE by itself fails to function, demonstrating that both factors act synergistically to activate h$MTII_A$ transcription. Finally, multiple nuclear factors interact with the mouse metallothionein MREs. Binding in vitro is stimulated 20- to 40-fold in the presence of heavy metals and correlates with subsequent activation of transcription (298). Genomic sequencing demonstrated that binding to five MREs within the rat metallothionein 1 promoter is induced in vivo by cadmium (299).

3.4 The β-interferon gene enhancer

Human α- and β-interferon (IFN) genes are usually silent, but they become quickly activated (within 1.5 h) in most if not all cell types, following viral infection or injection of double-stranded RNA (300). The level of induction depends on the cell line used and does not require de novo protein synthesis (300). Recent studies showed that poorly inducible cell lines like HeLa and 143 tk⁻ can be complemented when fused with highly induced cells (301). The level of induction also increases when poorly expressing cells are pre-treated with IFN and cyclohexamide, indicating that for efficient expression a factor is required which by itself can be induced by IFN.

The 5' flanking sequences contain regulatory elements for induction of expression both for α- (302) and β-interferons (303–305). According to some studies the region between positions −77 to −36 (relative to the cap site) is sufficient for the induction of β-interferon (304). This region, called the interferon regulatory element (IRE), has all the properties of an enhancer, inducible by viral infection or dsRNA (166,306). Other reports suggest that longer upstream sequences are required for regulation (306). The discrepancy probably results from differences between the cell lines used. Recently, it has been shown that tandem repeats of a 6 bp oligomer from the IRE can act as a virus-inducible enhancer (307). The corresponding region of the α-IFN gene (from positions −109 to −64 relative to the cap site) also confers inducibility on heterologous promoters (302,308).

Further analysis showed that the 3' part of the IRE contains repressing regulatory elements, while its 5' part is a constitutive transcriptional element responsible for positive regulation (309). Sequences between positions −210 to −107 also have a negative role, decreasing the basal level of expression without affecting inducibility (309). Genomic footprinting showed that before induction, the two repressor sites are occupied by proteins (310). After induction, these proteins dissociate and a new factor binds to the constitutive element. It appears then that under normal conditions, β-IFN gene expression is repressed by two proteins that bind in and around a constitutive transcriptional element. Upon viral infection or treatment with dsRNA, these proteins dissociate and binding of a positive factor leads to activation of transcription (311).

3.5 The c-*fos* enhancer

c-*fos* is a cellular oncogene that encodes a nuclear protein of unknown function. However, it has been demonstrated that c-*fos* is part of a protein complex that recognizes the AP-1 binding site and so probably regulates the expression of certain genes (312). c-*fos* is expressed at low levels in many cell types and at high levels in hematopoietic cells and extra-embryonal tissues (313,314). In cultured fibroblasts, expression of c-*fos* can be rapidly induced by growth factors. c-*fos* mRNA reaches maximum levels by 10–15 min after induction and then decreases abruptly (89).

Functional analysis of the 5' end of the gene showed that a short (~60 bp) fragment centered 300 bp upstream from the cap site has enhancer properties and is necessary and sufficient for inducibility of the c-*fos* (315,316). This area conforms to a DNase I hypersensitive site found in intact chromatin (316,317). Regulation of the c-*fos* gene requires additional elements since, for proper inducibility, sequences within the gene and at the 3' end are also necessary (315). Recent work shows that the gene is also under negative control (318).

A factor has been detected interacting with the c-*fos* enhancer (319,320). Interestingly, the binding of the factor is observed after stimulation by growth factors in A431 epidermal carcinoma cells (320) or after treatment of BALB/c-3T3 cells with conditioned medium from v-*sis*-transformed cells (321). The sequence of the binding site shows a dyad symmetry. Mutations which destroy symmetry also abolish binding and inducibility of the enhancer (322). This factor has been purified (323,324). Besides this factor, two more sites of DNA – protein interactions have been mapped between the c-*fos* enhancer and the cap site of the gene (325).

3.6 Enhancers in α-fetoprotein and other genes expressed in liver

α-Fetoproteins (AFP) are expressed at various levels in the visceral endoderm of the yolk sac and in the fetal liver and gastrointestinal tract (326). In F9 cells the AFP gene is activated after treatment with retinoic acid and subsequent differentiation to visceral endoderm (327). The sequences required for expression in a hepatoma cell line are scattered in an upstream region of 7 kb. Three different enhancer elements that stimulate transcription in liver, and to a lesser extent in HeLa cells, have been identified (328,329). Although the enhancer elements are tissue-specific, the major determinant of cell specificity is a promoter element between positions – 85 to – 52 upstream from the site of transcriptional initiation (328). The three enhancers I, II, III are equivalent and non-additive in the hepatoma cells, but they are not functionally redundant when tested in transgenic mice (330). The latter approach demonstrated that at least one enhancer is required for expression in all three tissues. In liver, enhancer I was slightly stronger than II and both were much stronger than III. In the gut, enhancers II and III were equivalent and much less active than I (330). This study also showed that the accessibility of the gene to *trans*-acting factors plays an important role in tissue-specific diversity of expression. Detailed analysis of the three enhancers showed that they are composed of positively and negatively acting domains (331).

It is worth mentioning that the three enhancer elements are located in the 14 kb intragenic region between the AFP and albumin genes. The two genes belong to the same family and although they diverged 300 – 500 million years ago (332), they are still tightly linked. It would be interesting to see if the intragenic elements are required for albumin gene expression as well, which in turn might explain the tight linkage of the two genes. Very recently though, an enhancer was identified in front of the albumin gene, 10 kb upstream from the cap site (333).

A tissue-specific enhancer element has been identified in prealbumin (transthyretin), another gene expressed in liver (334). The enhancer is located between 1.6 and 2.15 kb upstream from the cap site and interacts with both ubiquitous and liver-specific factors (335). Liver-specific enhancers have been also found in the human and mouse α_1-antitrypsin genes (336,337). A common liver-specific protein recognizes the enhancers of the transthyretin and α_1-antitrypsin genes (338).

3.7 Enhancers in the major histocompatibility genes

The major histocompatibility complex (MHC) class I genes are expressed in most cell types and are involved in several pathways of the immune system (339). Expression of these genes is induced by interferons (340). Endogenous and transfected MHC class I genes are not expressed in embryonal carcinoma cells, but become activated when these cells are induced to differentiate (341,342).

The H-2K^b MHC gene contains two enhancer elements called A and B found within 200 bp from the cap site (343). The corresponding homologous sequences of the MHC H-2L^d gene also display enhancer activity (344,345). Interestingly, enhancer sequences function poorly in undifferentiated EC cells (343). Moreover, the MHC enhancers repress activity from a heterologous promoter in F9 cells but stimulate activity in differentiated cells (344).

Enhancer A partially overlaps with sequences responsible for induction by interferons (343). In order for IFN-response sequences to function, they have to be linked to an enhancer (346). A nuclear factor called H2TF1 has been identified that binds to enhancer A (347). The binding site shows a perfect dyad symmetry with one copy overlapping the IFN-response sequence. This factor also binds to the β_2-microglobulin enhancer (348) and the SV40 core but with decreased affinity (347). NF-$\varkappa$B, the inducible B-cell specific factor, also interacts with this binding site, indicating that NF-$\varkappa$B and H2TF1 are related (349). H2TF1 (also called KBF1) has been purified (350). Its cDNA has been cloned employing a novel technique of screening cDNA expression libraries using the recognition site as a probe (351).

An enhancer has been characterized at the 5' region of the E$_\beta$ gene of the MHC class II genes in mouse (352). Sequences required for activity are found within a 2 kb area, about 600 bp upstream from the cap site. The enhancer is tissue-specific since it works in cells that produce class II antigens (i.e. B cells) and not in fibroblasts (352). Enhancer elements are also present in the human HLA-DQ genes (353).

3.8 Enhancers in prolactin and growth hormone genes

The anterior pituitary gland expresses a number of discrete trophic hormones. Among them, growth hormone is produced by somatotrophs and prolactin by lactotrophs. The two hormones have evolved from a common ancestral gene and in addition they exhibit a similar pattern of developmental regulation (see 354 for references).

In the case of the prolactin gene, a tissue-specific enhancer has been identified around 1.85 kb upstream from the cap site (354). The enhancer can be divided into two parts. Both are required for full activity, but their relative position and orientation is not important. A second enhancer element has been discovered within 80 bp from the cap site and is required for induction by epidermal growth factor (EGF) and TPA (355). This inducible enhancer closely resembles a similar element described above in the c-*fos* gene, but in the case of the prolactin gene only the 5′ half of the dyad symmetry is present.

A different cell-specific enhancer has been identified in the growth hormone gene, about 200 bp upstream from the cap site (354). A factor, present only in cells producing growth hormone, binds to the 3′ end of the enhancer. In disagreement with this report, another study showed that the enhancer is promiscuously active and growth hormone is expressed specifically in pituitary cells only when further upstream sequences are included (356). This finding suggests that these upstream sequences suppress expression in non-pituitary cells. As with other negative elements described so far, these sequences can repress transcription from heterologous promoters as well (356). Tissue-specific expression of the growth hormone gene is also regulated by cell-specific interaction of *trans*-acting factors with promoter sequences (357). Recently it was shown that a cell-specific factor activates both the growth hormone and prolactin genes (358).

3.9 Tissue-specific enhancers in other genes

An enhancer was found in the 5′ end of the proenkephalin gene (359). It consists of three closely related 12 bp sequences which are required for induction by phorbol esters and cAMP. A 43 kd protein (CREB) binds to the cAMP responsive element (CRE) following stimulation by cAMP (360). The placental-specific enhancer of the human glycoprotein hormone α-subunit gene also contains a CRE (361). The antithrombin III gene has a tissue-specific enhancer which is active in cells that produce this protein, namely liver and kidney (362). The enhancer shares extensive homology with the Ig$\varkappa$ chain enhancer. The chicken lysozyme gene contains an enhancer active in lysozyme-producing myeloid cells but not in chicken fibroblasts (363). Interestingly, the enhancer is located 6.1 kb upstream from the 5′ start site and coincides with a DNase I hypersensitive site observed only in lysozyme-producing cells. The mouse α-crystallin gene also has an enhancer, about 200 bp upstream from the cap site, which is active in lens cells and not in fibroblasts (364). A lens-specific enhancer has been found in the chicken δ_1-crystallin gene (365). The cytochrome P_1-450 gene has an enhancer which is induced by dioxin (366). Enhancers have been localized in the first introns of the human β-actin (367) and the mouse α_2-collagen genes (368). A muscle-specific enhancer is present about 1 kb upstream from the creatine kinase gene (369).

A temporal and tissue-specific enhancer has also been discovered in the chicken β-globin gene (370,371). Surprisingly, this element is located at the 3' end of the gene lying about 400 bp past the polyadenylation signal of the gene. Why the enhancer is located at the 3' end is not clear, but it might reflect evolution of the globin-gene cluster, or it might be required for regulation of other globin genes like the ε-embryonic gene that is nearby. Subsequently enhancers have been identified at the 3' end of the human β-globin (372,373) and ^Aγ-globin (374) genes, the duck β-globin (375) and the chicken histone H5 gene (376).

Enhancers have been found in *Drosophila* genes. Examples are the tissue-specific enhancers of the yolk protein 1 gene and the *Sgs-4* gene (377,378). Besides being necessary for expression in the fat body, the yolk protein enhancer determines sex- and developmental-stage-specific expression. A far upstream element has been found in the Fushi-tarazu gene with enhancer properties, although it has not yet been tested for 3' promoter activation (379). Enhancers in *Drosophila* are particularly important. The powerful tools available for research on this organism, such as its well established genetics, chromosomal mapping and P-element mediated transformation, might help resolve the mechanism of enhancer function.

In *Xenopus*, an enhancer has been found in the U2 gene (46). The enhancer includes the octamer motif found in the SV40 enhancer and the promoters and enhancers of immunoglobulin genes (see above). The U2 enhancer binds the same factor (84). In addition, a heat-inducible enhancer has been found in the heat shock gene hsp70 (380).

Enhancer-like elements have been found in yeast (381,382) and in plant genes (383,384), the latter being involved in light regulation and tissue-specificity. Neither the yeast nor plant elements stimulate transcription when placed at the 3' end of the assayed promoter.

4. Comparative analysis of enhancers

As information about cellular and viral enhancers accumulates, it becomes evident that enhancers are a rather heterologous group of *cis*-acting elements. First, there is a large variation in their size which can range from 40 to 60 bp as in the case of BPV and c-*fos* enhancers, to about 700 bp for mouse cytomegalovirus enhancer. Second, their location relative to the gene is highly variable. Usually enhancers are found at the 5' end of the gene, about 100 bp from the cap site (SV40, insulin, interferons etc.) but they can be as far as 6, 7 or even 10 kb upstream from the gene (lysozyme, fetoprotein, albumin) or at the 3' end past the polyadenylation site (β-globin, BPV), within an intron (immunoglobulins) or even in the middle of a coding region (hepatitis B). Third, enhancers do not show any extensive sequence homology. This heterogeneity demonstrates that in order to test if a certain gene is under the influence of an enhancer, one might have to functionally analyze different subfragments covering large areas on both sides of the transcriptional unit.

Enhancers also display different modes of action. For example, some of them are constitutively active in many cell types (SV40). Others are cell-specific and even work in particular stages during development (immunoglobulin, insulin, lysozyme, β-globin, yolk protein etc.). Others stimulate transcription when induced by a variety of agents. For instance, the interferon enhancer is induced by viral infection, c-*fos* by growth factors, MTV by glucocorticoids, hsp70 enhancer by heat. Some enhancers are both constitutively active and inducible at the same time. Examples are the MSV and the metallothionein enhancers.

Alternative approaches have been designed to identify unknown enhancers within eukaryotic genomes. These include screening genomic libraries with viral enhancers as probes (138,385), applying selective pressure (386–388), randomly isolating activator elements from the genome (389), or using an 'enhancer trap' (53). These approaches have generally been met with limited success because it is difficult to assign a function to the isolated sequences. The 'trap' proved extremely useful though for screening subfragments of large viral genomes like herpes viruses for the presence of enhancers (191–193). Moreover, creation of functional enhancers by rearranging sequences at the ends of the 'trap' have led to a better understanding of the nature of enhancers.

When analyzing data for enhancer elements, one should also keep in mind that the results might depend on the conditions of the assay system used. For example, the mapped borders of the β-IFN enhancer differ depending on the cell line used (304,306). The three enhancer elements of the α-fetoprotein gene appear equivalent when tested in a hepatoma cell line, but they display different relative strengths when analyzed in transgenic mice (328,330). The effect of SV40 enhancer on certain promoters like HSV-TK varies according to the assay system (26). In addition, some promoters are resistant to activation by a particular enhancer (24,25,390).

Despite this diversity, enhancers have two common characteristics: an exceptional positional flexibility and a modular type of structure (391–393). Mutational analysis showed that enhancers consist of a plethora of short sequence motifs. Although the full set of these motifs is required for optimal activity, the enhancer can still function even if a large number of them is deleted. It is obvious then that enhancers contain redundant information and, as a result, they are tolerant of a wide range of sequence manipulations (391). This is particularly true for viral enhancers which need to compete with and overcome cellular genes in order to use the transcriptional machinery to their advantage. Recent analysis shows that cellular enhancers like the immunoglobulin gene enhancer also have a modular organization (253–255).

In some cases, and again this is particularly true for viral enhancers such as in SV40, BKV, LPV, JCV and cytomegaloviruses, enhancers contain duplications. Why duplicate a particular sequence? A possible answer is provided by studies with the SV40 enhancer. It was shown that a certain enhancer unit (e.g. the core or octamer motif) by itself is not sufficient for activation. When this motif is duplicated though, it shows enhancing activity (394,395). The activity increases with the number of copies used to construct these artificial enhancers. The same is true for the whole enhancer as multiplication of the 72 bp repeats leads to a linear increase with copy number (21). Moreover, using multiple copies of GRE also leads to an additive increase in transcription (396). The clear conclusion is that by repeating a certain number of motifs, the potential of the enhancer to activate transcription increases. But even when repeats are not present, the redundancy of information still exists.

The oligomerization of single enhancer motifs also showed that the different subfragments of the SV40 enhancer have distinct cell specificities (394,395). Remarkably, the core motif has a pattern of activity similar to the entire enhancer. The Pvu II motif or 'pseudocore' is active only in CV-1 cells, while the Sph I motifs which include the octamer are active in lymphoid cells. Presumably, this differential activity reflects an analogous distribution of the corresponding *trans*-acting factors in the various cell types (see also discussion below).

5. How do enhancers work?

Many efforts have been devoted to enhancers and their properties. It is known that enhancers increase the number of RNA polymerase molecules in a neighboring gene (22,23), but still the mechanism by which this is accomplished is not well understood. This uncertainty is reflected in the number of models that have been proposed to explain how enhancers work. Generally, they can be divided into two groups: models which imply an indirect effect of the enhancer on transcriptional activation and models that implicate the enhancer directly in transcriptional stimulation.

5.1 Indirect effects of enhancers in transcriptional stimulation

5.1.1 Enhancers as inducers of DNA conformational changes

According to this model, enhancers induce DNA conformational changes which can then stimulate transcription. These changes might include unwinding of DNA or creation of a new form (e.g. Z-DNA). Such alternative structures could stabilize a transcriptional complex that forms at the initiation site of transcription. The role of *trans*-acting factors could then be to induce and maintain these new forms of DNA.

Evidence for such a model is provided by the fact that torsional stress stimulates transcription (397,398). Moreover, topoisomerase II sites have been mapped in and around the SV40 enhancer (399). Evidence against such a model comes from experiments where the enhancer was topologically separated from the promoter in 'tailed circle' plasmids (400). In this case, torsional changes could not be induced in the area of the promoter. Nevertheless, the enhancer was still able to stimulate transcription. In addition, it has been shown that different parts of activator proteins are responsible for binding and transcriptional activation, or in other words, binding only is not sufficient for *trans*-activation (401,402). It is also difficult to conceive how structural changes could be propagated over distances of 10 000 bp.

5.1.2 Enhancers as inducers of chromatin structure changes

In this model, enhancers induce changes in chromatin structure which in turn render the corresponding gene more accessible to RNA polymerase and the transcriptional machinery in general. This model is supported by the fact that DNase I hypersensitive sites have been mapped to enhancer elements such as SV40, polyoma and immunoglobulin (61,243,403). Furthermore, the region in and around the SV40 enhancer appears as a nucleosomal gap in electron micrographs of the viral minichromosomes (62). This is an intrinsic property of the enhancer sequences since moving the enhancer to new sites in the viral genome also relocates the gap and using two copies of the enhancer increases the length of the gap respectively. It is possible, though, that these enhancer properties simply reflect the binding of *trans*-acting factors rather than an enhancer stimulated conformational change in chromatin structure. Besides, enhancers work to a limited degree *in vitro* using purified DNA templates and nuclear extracts, so to some extent chromatin structure is not important.

5.1.3 The nuclear address model

According to this model, enhancers act as guides in directing genes into compartments of the nucleus rich in RNA polymerase and other transcriptional factors. This can be accomplished if the enhancer is responsible for association of the gene to specific sites in the nuclear scaffold or the nuclear matrix. It is true that actively transcribed SV40 chromosomes as well as other active genes are associated with the nuclear matrix (404–408). It is interesting that scaffold attachment regions (SAR) have been found close to the enhancer/promoter elements of three *Drosophila* genes (409) and the immunoglobulin κ chain enhancer (410). This cohabitation of regulatory elements and SAR is certainly intriguing and it might prove that transcriptional enhancement is a very complex phenomenon.

5.2 Direct involvement of enhancers in transcription activation

5.2.1 Enhancers are the targets of transcriptional factors

Our description of the various viral and cellular enhancers pinpoints the fact that enhancers have a modular structure composed of many short sequence motifs. The role of these motifs is to bind cellular or in some cases viral proteins (411). We have already discussed in previous sections examples that correlate the binding of these factors with enhancer activity and stimulation of transcription. Briefly, evidence is provided by the following types of experiments.

(i)   Competition assays both *in vivo* and *in vitro* showed that using increasing amounts of enhancer sequences as competitor DNA abolishes activity (57 – 59,259).

(ii)   Mutations in the motifs that decrease or eliminate activity also abolish the binding of the corresponding factor, as was shown for SV40 (64) and c-*fos* enhancers (322). Moreover, mutants deficient in activity also fail to compete for *trans*-acting factors in competition experiments (59).

(iii) Enhancer activity correlates with the binding of factors to crucial sequences. Examples are the GRE and the glucocorticoid receptor (202,204), AP-1 binding and activation of h$MTII_A$ basal level enhancer (68), the binding to the octamer motif and stimulation of transcription (84), the binding of factors to the insulin enhancer only in insulin producing cells (293), the binding of proteins to the MRE after metal induction (298), the binding of factors to the β-IFN constitutive enhancer after dsRNA induction (310) and the binding of factors to the c-*fos* enhancer after growth factor stimulation (320).

A very plausible model then is that these proteins are involved directly in the formation of transcriptional complexes, which in turn will render enhancers entry sites for *trans*-acting transcriptional factors. The situation will be analogous to prokaryotes which have operator sites that require the binding of repressor and activator molecules to modulate transcription. In order to provide solid support for this model, the formation of such complexes has to be demonstrated in appropriate *in vitro* systems.

Along these lines, one could argue that there are no intrinsic differences between enhancers and promoters, but all these control elements are simply binding sites for transcriptional factors. The differences we then observe in their location and positional flexibility is simply a reflection of the relative location of the corresponding proteins within the transcriptional complex. For example, a TATA binding protein together with other promoter binding factors might provide the basis for such a complex. These proteins have to be within a certain distance from the cap site in order for them to be able to participate in the complex. Some freedom exists, since some of these proteins can still act even if their binding site is placed in opposite orientation (13,14). The enhancer binding proteins could occupy positions within the complex that allow flexibility in the location of their binding sites at the DNA level.

How can this be accomplished? There are two ways to solve the problem. Either these proteins can slide from their original binding site within the enhancer to the transcriptional site of initiation, or alternatively the proteins can be brought in direct contact with the transcriptional machinery by looping out the intervening DNA (412). The fact that proteins binding to enhancers make contact with promoter binding proteins has been elegantly demonstrated in the case of the SV40 elements (30). The results are analogous with similar experiments using the repressor and its operator site in prokaryotes. In this case, it was shown that cooperative binding, which involves interaction of two repressor molecules, can take place only if the two binding sites are separated by integral numbers of helical turns, or in other words, the two molecules have to be on the same side of the helix (413). The interaction of the two repressors, with the intervening DNA looping out, was visualized directly under the electron microscope (414). Formation of these loop structures was also implied in other prokaryotic systems in order to explain regulation of transcription from distal sites (*gal* and *ara*BAD operons; 415,416). Although a sliding model would explain the proximal promoter activation observed in some cases, so can the looping model since statistically it might be easier to form loops with the most proximal site available. But sliding or other tracking mechanisms cannot explain why inserting 5 bp between the SV40 enhancer and its promoter (or otherwise between the Sp1 and AP-1 proteins) decreases activity more than inserting 10 bp (30).

It is appropriate to emphasize that the enhancer might be required only for establishment of these transcriptional complexes and not for their maintenance since it appears that the enhancer is dispensable when transcription starts (see ref. 281 and references therein).

5.2.2 *Trans-acting factors might play multiple roles*

Recently, some of the factors interacting with promoter and enhancer sequences have been isolated. From some of these data, it appears that different factors might bind to the same sequence motif. For example, two distinct proteins have been isolated and shown to interact with the CCAAT box (417,418), and probably several others exist (419,420). One of them proved to be indistinguishable for NFI (111). It is also possible that the same protein interacts with different sequence motifs (83,421).

Certain factors may play opposite roles in different cell-types. It is known, for example, that elimination of the μE1 site from the immunoglobulin enhancer reduces its activity in B cells (253,254). At the same time, elimination of μE1 allows the enhancer to work in fibroblasts (253,254). It is possible that two different proteins bind to the same site, one a repressor present in fibroblasts, the other an activator present in lymphoid cells. Alternatively, the same protein is present in both cell types. In fibroblasts it represses transcription because the combination of factors is such that the protein hinders formation of an active transcriptional complex. In B cells, the same protein stimulates transcription because additional factors are present that allow its participation in the transcriptional complex.

A recent analysis shows that the octamer motif is present in the promoter of the HSV-TK gene (390). In the absence of an enhancer, the octamer supports a higher rate of transcription. When the promoter is under the influence of the $\varkappa$ chain enhancer, again the octamer acts positively. But when the tk promoter is under the influence of the MSV enhancer, the presence of the octamer in the promoter area prevents enhancement. These results could be explained with similar arguments as above. That is, in one case the combination of factors is such that the octamer binding protein(s) participate in a transcriptional complex, while in the other case the same protein(s) might hinder formation of a complex if the combination of factors is unfavorable. As was discussed earlier, these results demonstrate once more that the interaction between enhancers and promoters, presumably via *trans*-acting factors, is important for transcription stimulation.

6. Summary and Perspectives

Multiple levels of control regulate the flow of genetic information in eukaryotic organisms. A first general type of regulation is exerted by the chromatin structure itself (see ref. 422 and references therein). It has been known for some time that genes which are not expressed in certain tissues are packaged in heterochromatin and they are no longer accessible to transcriptional factors. In this way, genes that encode products of terminally differentiated cells like globins and immunoglobulins might be kept silent in unrelated tissues. In addition, some DNA sequences may help keep genes inactive, as might be the case of the repetitive sequence LINES found in the proximity of the insulin gene (292).

Recently it has been shown that position-independent expression of β-globin genes in transgenic mice could be achieved if sequences 50 kb 5' and 20 kb 3' of the gene were included in the constructs (423). These results demonstrate that certain sequences might regulate the tissue-specific accessibility of a large chromatin locus. When the locus is open, the tissue- and/or stage-specific expression of genes is accomplished by a combination of positively- and negatively-acting transcriptional factors. The promoter and enhancer regions contain a certain number of binding sites. An enormous number of different combinations can be created with only a small number of distinct *cis*-acting motifs and their corresponding *trans*-acting factors. Consequently, only a unique combination of binding sites for transcription factors is required for specific expression, rather than unique factors for every single gene. This way a gene will be expressed in cells that contain a proper combination of *trans*-acting factors to interact with sequence motifs present in its regulatory area. The gene will be inactive in tissues where the corresponding factors are missing or if it has no motifs for transcriptional factors present in a certain tissue.

The combinatorial effects discussed above can be expanded if one includes repressor sites among the transcriptional elements. This type of control might apply to genes that are, under normal conditions, inactive but are quickly stimulated following an external signal. A characteristic example is the interferon gene where repressor sites are occupied when the gene is not expressed. The bound proteins dissociate after induction and thus allow the binding of factors to the constitutive enhancer.

Some of the factors might be present in a cell type all the time while others might be induced by certain signals (growth factors, hormones, infection etc). In this case either the external stimulus induces expression of the factor or the factor is activated by some post-translational modification. An example may be proteins binding to the c-*fos* enhancer after stimulation by growth factors. Possibly one of the effects mediated by protein kinase C is the phosphorylation and activation of some transcriptional factors.

What lies ahead? Certainly more information will become available about *cis*-acting regulatory elements of eukaryotic genes. In addition, *trans*-acting factors will be isolated, purified and characterized. There is a need to test the function of these factors in appropriate *in vitro* systems as in the case of AP-1. It will also be necessary to study precisely how these proteins enhance transcription, in particular a direct demonstration that they are involved in the formation of transcriptional complexes or in some other function such as the induction of DNA structural changes. This will be accomplished only when the full range of factors interacting with a gene are isolated and the various regulatory events can be reproduced *in vitro*. A major step forward will be made when the genes coding for *trans*-acting factors themselves are isolated. Mutations in their coding regions will allow us to dissect their different functional parts and thus reveal their mode of action. Study of the regulation of expression of *trans*-acting factor genes might provide clues about molecular events that lead to differentiation. This is particularly true for factors found only in some cell types. It will be interesting to see how the expression and regulation of their genes are linked to events related to cell differentiation and development.

7. Acknowledgements

We would like to thank A.Püschel, P.Duprey and J.Erselius for helpful comments during the preparation of the manuscript. We are indebted to U.Drescher for his help in preparing the figures and to Rosemary Drescher for typing parts of the manuscript. We would also like to thank many of our colleagues for providing reprints and manuscripts prior to publication. A.H. is a recipient of a long-term EMBO Fellowship. This work was supported by grant BCT 0364/0 from the Bundesministerium für Forschung und Technologie to P.G. U.S. was supported by Fonds der Chemischen Industrie.

8. References

1. Ptashne,M. (1986) *A Genetic Switch.* Cell Press and Blackwell Scientific Publications, Cambridge, Palo Alto.
2. Breathnach,R. and Chambon,P. (1981) Organization and expression of eukaryotic split genes coding for proteins. *Annu. Rev. Biochem., 50*, 349.
3. McKnight,S.L. and Kingsbury,R.C. (1982) Transcriptional control signals of a eukaryotic protein-coding gene. *Science, 217*, 316.
4. McKnight,S.L. and Tjian,R. (1986) Transcriptional selectivity of viral genes in mammalian cells. *Cell, 46*, 795.
5. Grosschedl,R. and Birnstiel,M.L. (1980) Spacer DNA sequences upstream of the TATAAATA sequence are essential for promotion of H2A histone gene transcription *in vivo. Proc. Natl. Acad. Sci. USA, 77*, 7102.
6. Benoist,C. and Chambon,P. (1981) *In vivo* sequence requirements of the SV40 early promoter region. *Nature, 290*, 304.
7. Gruss,P., Dhar,R. and Khoury,G. (1981) Simian virus 40 tandem repeated sequences as an element of the early promoter. *Proc. Natl. Acad. Sci. USA, 78*, 943.
8. Tooze,J. (ed.) (1980) *DNA Tumor Viruses.* Cold Spring Harbor Laboratory, Cold Spring Harbor, NY.
9. Moreau,P., Hen,R., Wasylyk,B., Everett,R., Gaub,M.P. and Chambon,P. (1981) The SV40 72 base pair repeat has a striking effect on gene expression both in SV40 and other chimeric recombinants. *Nucleic Acids Res., 9*, 6047.
10. Banerji,J., Rusconi,S. and Schaffner,W. (1981) Expression of a β-globin gene is enhanced by remote SV40 DNA sequences. *Cell, 27*, 299.
11. Fromm,M. and Berg,P. (1983) Simian virus 40 early- and late-region promoter functions are enhanced by the 72 base pair repeat inserted at distant locations and inverted orientations. *Mol. Cell. Biol., 3*, 991.
12. Capecchi,M.R. (1980) High efficiency transformation by direct micro-injection of DNA into cultured mammalian cells. *Cell, 22*, 479.
13. Everret,R.D., Baty,D. and Chambon,P. (1983) The repeated GC-rich motifs upstream from the TATA box are important elements of the SV40 early promoter. *Nucleic Acids Res., 11*, 2447.
14. Gidoni,D., Kadonaga,J.T., Barrera-Saldana,H., Takahashi,K., Chambon,P. and Tjian,R. (1985) Bidirectional SV40 transcription mediated by tandem Sp1 binding interactions. *Science, 230*, 511.
15. Picard,D. (1985) Viral and cellular transcription enhancers. *Oxford Surveys on Eukaryotic Genes, 2*, 24.
16. Neuhaus,G., Neuhaus-Url,G., Gruss,P. and Schweiger,H.G. (1984) Enhancer-controlled expression of the simian virus 40 T-antigen in the green alga *Acetabularia. EMBO J., 3*, 2169.
17. Laimins,L.A., Khoury,G., Gorman,C., Howard,B. and Gruss,P. (1982) Host-specific activation of transcription by tandem repeats from simian virus 40 and Moloney murine sarcoma virus. *Proc. Natl. Acad. Sci. USA, 79*, 6453.
18. de Villiers,J., Olson,L., Tyndall,C. and Schaffner,W. (1982) Transcriptional 'enhancers' from SV40 and polyoma virus show a cell type preference. *Nucleic Acids Res., 10*, 7965.
19. Byrne,B.J., Davis,M.S., Yamaguchi,J., Bergsma,D.J. and Subramanian,K.N. (1983) Definition of the simian virus 40 early promoter region and demonstration of a host range bias in the enhancement effect of the simian virus 40 72-base-pair repeat. *Proc. Natl. Acad. Sci. USA, 80*, 721.

20. Spandidos,D.A. and Wilkie,N.M. (1983) Host-specificities of papilloma-virus, Moloney murine sarcoma virus and simian virus 40 enhancer sequences. *EMBO J.*, **2**, 1193.
21. Zenke,M., Grundström,T., Matthes,H., Wintzerith,M., Schatz,C., Wildeman,A. and Chambon,P. (1986) Multiple sequence motifs are involved in SV40 enhancer function. *EMBO J.*, **5**, 387.
22. Treisman,R. and Maniatis,T. (1985) Simian virus 40 enhancer increases number of RNA polymerase II molecules on linked DNA. *Nature*, **315**, 72.
23. Weber,F. and Schaffner,W. (1985) Simian virus 40 enhancer increases RNA polymerase density within the linked gene. *Nature*, **315**, 75.
24. Humphries,R.K., Ley,T., Turner,P., Moulton,A.D. and Nienhuis,A.W. (1982) Differences in human α-, β- and δ-globin gene expression in monkey kidney cells. *Cell*, **30**, 173.
25. Treisman,R., Green,M.R. and Maniatis,T. (1983) *Cis* and *trans* activation of globin gene transcription in transient assays. *Proc. Natl. Acad. Sci. USA*, **80**, 7428.
26. Robbins,P.D., Rio,D.C. and Botchan,M.R. (1986) *Trans* activation of the simian virus 40 enhancer. *Mol. Cell. Biol.*, **6**, 1283.
27. Mason,J.O., Williams,G.T. and Neuberger,M.S. (1985) Transcription cell type specificity is conferred by an immunoglobulin V_H gene promoter that includes a functional consensus sequence. *Cell*, **41**, 479.
28. Wasylyk,B., Wasylyk,C. and Chambon,P. (1984) Short and long range activation by the SV40 enhancer. *Nucleic Acids Res.*, **12**, 5589.
29. Picard,D. and Schaffner,W. (1983) Correct transcription of a cloned mouse immunoglobulin gene *in vivo*. *Proc. Natl. Acad. Sci. USA*, **80**, 417.
30. Takahashi,K., Vigneron,M., Matthes,H., Wildeman,A., Zenke,M. and Chambon,P. (1986) Requirement of stereospecific alignments for initiation from the simian virus 40 early promoter. *Nature*, **319**, 121.
31. Wasylyk,B., Wasylyk,C., Augereau,P. and Chambon,P. (1983) The SV40 72 bp repeat preferentially potentiates transcription starting from proximal natural or substitute promoter elements. *Cell*, **32**, 503.
32. Kadesch,T. and Berg,P. (1986) Effect of the position of the simian virus 40 enhancer on expression of multiple transcription units in a single plasmid. *Mol. Cell. Biol.*, **6**, 2593.
33. Atchinson,M.L. and Perry,R.P. (1986) Tandem kappa immunoglobulin promoters are equally active in the presence of the kappa enhancer: implications for models of enhancer function. *Cell*, **46**, 253.
34. Herr,W. and Gluzman,Y. (1985) Duplications of a mutated simian virus 40 enhancer restore its activity. *Nature*, **313**, 711.
35. Khoury,G. and Gruss,P. (1983) Enhancer elements. *Cell*, **33**, 313.
36. Gruss,P. (1984) Magic enhancers? *DNA*, **3**, 1.
37. Schlokat,U. and Gruss,P. (1986) Enhancers as control elements for tissue-specific transcription. In *Oncogenes and Growth Control*. Kahn,P. and Graf,T. (eds), Springer-Verlag, Berlin, p.226.
38. Gluzman,Y. (ed.) (1985) *Eukaryotic Transcription: the Role of Cis- and Trans-acting Elements in Initiation*. Cold Spring Harbor Laboratory, Cold Spring Harbor, NY.
39. Weiher,H., König,M. and Gruss,P. (1983) Multiple point mutations affecting the simian virus 40 enhancer. *Science*, **219**, 626.
40. Falkner,F.G. and Zachau,H.G. (1984) Correct transcription of an immunoglobulin $\varkappa$ gene requires an upstream fragment containing conserved sequence elements *Nature*, **310**, 71.
41. Parslow,T.G., Blair,D.L., Murphy,W.J. and Granner,D.K. (1984) Structure of the 5'-ends of immunoglobulin genes; a novel conserved sequence. *Proc. Natl. Acad. Sci. USA*, **81**, 2650.
42. Bergman,Y., Rice,D., Grosschedl,R. and Baltimore,D. (1984) Two regulatory elements for $\varkappa$ immunoglobulin gene expression. *Proc. Natl. Acad. Sci. USA*, **81**, 7041.
43. Krol,A., Lund,E. and Dahlberg,J.E. (1985) The two embryonic U1 RNA genes of *Xenopus laevis* have both common and gene-specific transcription signals. *EMBO J.*, **4**, 1529.
44. Ciliberto,G., Buckland,R., Cortese,R. and Philipson,L. (1985) Transcription signals in embryonic *Xenopus laevis* U1 RNA genes. *EMBO J.*, **4**, 1537.

45. Ares,M.,Jr, Mangin,M. and Weiner,A.M. (1985) Orientation-dependent transcriptional activator upstream of a human U2 snRNA gene. *Mol. Cell. Biol.*, **5**, 1561.
46. Mattaj,I.W., Lienhard,S., Jiricny,J. and De Robertis,E.M. (1985) An enhancer-like sequence within the *Xenopus* U2 gene promoter facilitates the formation of stable transcription complexes. *Nature,* **316**, 163.
47. Falkner,F.G., Mocikat,R. and Zachau,H.G. (1986) Sequences closely related to an immunoglobulin gene promoter/enhancer element occur also upstream of other eukaryotic and prokaryotic genes. *Nucleic Acids Res.,* **14**, 8819.
48. Herr,W. and Clarke,J. (1986) The SV40 enhancer is composed of multiple functional elements that can compensate for one another. *Cell,* **45**, 461.
49. Veldman,G.M., Lupton,S. and Kamen,R. (1985) Polyomavirus enhancer contains multiple redundant sequence elements that activate both DNA replication and gene expression. *Mol. Cell. Biol.,* **5**, 649.
50. Herbomel,P. Bourachot,B. and Yaniv,M. (1984) Two distinct enhancers with different cell specificities coexist in the regulatory region of polyoma. *Cell,* **39**, 653.
51. Sen,R. and Baltimore,D. (1986) Multiple nuclear factors interact with the immunoglobulin enhancer sequences. *Cell,* **46**, 705.
52. Nordheim,A. and Rich,A. (1983) Negatively supercoiled simian virus 40 DNA contains Z-DNA segments within transcriptional enhancer sequences. *Nature,* **303**, 674.
53. Weber,R., de Villiers,J. and Schaffner,W. (1984) An SV40 'enhancer trap' incorporates exogenous enhancers or generates enhancers from its own sequences. *Cell,* **36**, 983.
54. Swimmer,C. and Shenk,T. (1984) A viable simian virus 40 variant that carries a newly generated sequence reiteration in place of the normal duplicated enhancer element. *Proc. Natl. Acad. Sci. USA,* **81**, 6652.
55. Petterson,M. and Schaffner,W. (1987) A purine-rich DNA sequence motif present in SV40 and lymphotropic papovarius binds a lymphoid-specific factor and contributes to enhancer activity in lymphoid cells. *Genes and Development,* **1**, 962.
56. Sassone-Corsi,P., Dougherty,J.P., Wasylyk,B. and Chambon,P. (1984) Stimulation of *in vitro* transcription from heterologous promoters by the simian virus 40 enhancer. *Proc. Natl. Acad. Sci. USA,* **81**, 308.
57. Sergeant,A., Bohmann,D., Zentgraf,H., Weiher,H. and Keller,W. (1984) A transcription enhancer acts *in vitro* over distances of hundreds of base-pairs on both circular and linear templates but not on chromatin-reconstituted DNA. *J. Mol. Biol.,* **180**, 577.
58. Wildeman,A.G., Sassone-Corsi,P., Grundström,T., Zenke,M. and Chambon,P. (1984) Stimulation of *in vitro* transcription from the SV40 early promoter by the enhancer involves a specific *trans*-acting factor. *EMBO J.,* **3**, 3129.
59. Schöler,H.R. and Gruss,P. (1984) Specific interaction between enhancer-containing molecules and cellular components. *Cell,* **36**, 403.
60. Sassone-Corsi,P., Wildeman,A. and Chambon,P. (1985) A *trans*-acting factor is responsible for the simian virus 40 enhancer activity *in vitro*. *Nature,* **313**, 458.
61. Cremisi,C. (1981) The appearance of DNase I hypersensitive sites at the 5' end of the late SV40 genes is correlated with the transcriptional switch. *Nucleic Acids Res.,* **9**, 5949.
62. Jongstra,J., Reudelhuber,T.L., Oudet,P., Benoist,C., Chae,C.-B., Jeltsch,J.-M., Mathis,D.J. and Chambon,P. (1984) Induction of altered chromatin structures by simian virus 40 enhancer and promoter elements. *Nature,* **307**, 708.
63. Wildeman,A.G., Zenke,M., Schatz,C., Wintzerith,M., Grundström,T., Matthes,H., Takahashi,K. and Chambon,P. (1986) Specific protein binding to the simian virus 40 enhancer *in vitro*. *Mol. Cell. Biol.,* **6**, 2098.
64. Davidson,I., Fromental,C., Augereau,P., Wildeman,A., Zenke,M., Chambon,P. (1986) Cell-type specific protein binding to the enhancer of simian virus 40 in nuclear extracts. *Nature,* **323**, 544.
65. Rosales,R., Vigneron,M., Macchi,M., Davidson,I., Xiao,J.H. and Chambon,P. (1987) *In vitro* binding of cell-specific and ubiquitous nuclear proteins to the octamer motif of SV40 enhancer and related motifs present in other promoters and enhancers. *EMBO J.,* **6**, 3015.
66. Xiao,J.H., Davidson,I., Ferrandon,D., Rosales,R., Vigneron,M., Macchi,M., Ruffenach,F. and Chambon,P. (1987) One cell-specific and three ubiquitous nuclear

proteins bind *in vitro* to overlapping motifs in the domain B1 of the SV40 enhancer. *EMBO J.*, **6**, 3005.

67. Xiao,J.H., Davidson,I., Macchi,M., Rosales,R., Vigneron,M., Staub,A. and Chambon,P. (1987) *In vitro* binding of several cell-specific and ubiquitous nuclear proteins to the GT-I motif of the SV40 enhancer. *Genes and Development*, **1**, 794.

68. Lee,W., Haslinger,A., Karin,M and Tjian,R. (1987) Activation of transcription by two factors that bind promoter and enhancer sequences of the human metallothionein gene and SV40. *Nature*, **325** 368.

69. Dynan,W.S. and Tjian,R. (1983) Isolation of transcription factors that discriminate between different promoters recognized by RNA polymerase II. *Cell*, **32**, 669.

70. Dynan,W.S. and Tjian,R. (1983) The promoter-specific transcription factor Sp1 binds to upstream sequences in the SV40 early promoter. *Cell*, **32** 79.

71. Gidoni,D., Dynan,W.S. and Tjian,R. (1984) Multiple specific contacts between a mammalian transcription factor and its cognate promoters. *Nature*, **312**, 409.

72. Dynan,W.S., Saffer,J.D., Lee,W. and Tjian,R. (1985) Transcription factor Sp1 recognizes promoter sequences from the monkey genome that are similar to the simian virus 40 promoter. *Proc. Natl. Acad. Sci. USA*, **82**, 4915.

73. Dynan,W.S., Sazer,S., Tjian,R. and Schimke,R.T. (1986) The transcription factor SO1 recognizes a DNA sequence in mouse dihydrofolate reductase promoter. *Nature*, **319**, 246.

74. Kadonaga,J.T. and Tjian,R. (1986) Affinity purification of sequence-specific DNA binding proteins. *Proc. Natl. Acad. Sci. USA*, **83**, 5889.

75. Kadonaga,J.T., Carner,K.R., Masiarz,F.R. and Tjian,R. (1987) Isolation of cDNA encoding transcription factor Sp1 and functional analysis of the DNA binding domain. *Cell*, **51**, 1079.

76. Bos,T.J., Bohman,D., Tsuchie,H., Tjian,R. and Vogt,P.K. (1988) v-*jun* encodes a nuclear protein with enhancer binding properties of AP-1. *Cell*, **52**, 705.

77. Bohman,D., Bos,T.J., Admon,A., Nishimura,T., Vogt,P.K. and Tjian,R. (1987) Human proto-oncogene c-*jun* encodes a DNA binding protein with structural and functional properties of transcriptional factor AP-1. *Science*, **238**, 1386.

78. Angel,P., Allegretto,E.A., Okino,S.T., Hattori,K., Boyle,W.J., Hunter,T. and Karin,M. (1988) Oncogene *jun* encodes a sequence-specific *trans*-activator similar to AP-1. *Nature*, **332**, 166.

79. Vogt,P.K., Bos,T.J. and Doolittle,R.F. (1987) Homology between the DNA-binding domain of the GCN4 regulatory protein of yeast and the carboxyl-terminal region of a protein coded for by the oncogene *jun*. *Proc. Natl. Acad. Sci. USA*, **84**, 3316.

80. Struhl,K. (1987) The DNA-binding domains of the *jun* oncoprotein and the yeast GCN4 transcriptional activator protein are functionally homologous. *Cell*, **50**, 841.

81. Struhl,K. (1988) The *jun* oncoprotein, a vertebrate transcription factor, activates transcription in yeast. *Nature*, **332**, 649.

82. Mitchell,P.J., Wang,C. and Tjian,R. (1987) Positive and negative regulation of transcription *in vitro*: enhancer binding protein AP-2 is inhibited by SV40 T antigen. *Cell*, **50**, 847.

83. Johnson,P.F., Landschulz,W.H., Graves,B.J. and McKnight,S.L. (1987) Identification of a rat liver nuclear protein that binds to the enhancer core element of three animal viruses. *Genes and Development*, **1**, 133.

84. Bohman,D., Keller,W., Dale,T., Schöler,H.R., Tebb,G. and Mattaj,I.W. (1987) A transcription factor which binds to the enhancers of SV40, immunoglobulin heavy chain and U2 snRNA genes. *Nature*, **325**, 268.

85. Sturm,R., Baumruker,T., Franza,R.B.,Jr and Herr,W. (1987) A 100-kd HeLa cell octamer binding protein (OBP100) interacts differently with two separate octamer-related sequences within the SV40 enhancer. *Genes and Development*, **1**, 1147.

86. Fletcher,C., Heinz,N. and Roeder,R.G. (1987) Purification and characterization of OTF-1, a transcription factor regulating cell cycle expression of a human histone H2b gene. *Cell*, **51**, 773.

87. Scheidereit,C., Heguy,A. and Roeder,R.G. (1987) Identification and purification of a human lymphoid-specific octamer-binding protein (OTF-2) that activates transcription of an immunoglobulin promoter *in vitro*. *Cell*, **51**, 783.

88. Imbra,R.J. and Karin,M. (1986) Phorbol ester induces the transcriptional stimulatory activity of the SV40 enhancer. *Nature*, **323**, 555.

89. Greenberg,M.E. and Ziff,E.B. (1984) Stimulation of 3T3 cells induces transcription of the c-*fos* proto-oncogene. *Nature,* **311**, 433.
90. Chiu,R., Imagawa,M., Imbra,R.I., Bockoven,J.R. and Karin,M. (1987) Multiple *cis*- and *trans*-acting elements mediate the transcriptional response to phorbol esters. *Nature,* **329**, 648.
91. Imagawa,M., Chiu,R. and Karin,M. (1987) Transcription factor AP-2 mediates induction by two different signal-transduction pathways: protein kinase C and cAMP. *Cell,* **51**, 251.
92. Botchan,M., Grodzicker,T. and Sharp,P.A. (eds) (1986) *DNA Tumor Viruses: Control of Gene Expression and Replication.* Cold Spring Harbor Laboratory, Cold Spring Harbor, NY.
93. Tyndall,C., La Mantia,G., Thacker,C.M., Favaloro,J. and Kamen,R. (1981) A region of the polyoma virus genome between the replication origin and late protein coding sequences is required in *cis* for both early gene expression and viral DNA replication. *Nucleic Acids Res.,* **9**, 6231.
94. de Villiers,J. and Schaffner,W. (1981) A small segment of polyoma virus DNA enhances the expression of a cloned β-globin gene over a distance of 1400 base pairs. *Nucleic Acids Res.,* **9**, 6251.
95. Soeda,E., Arrand,J.R., Smolar,N. and Griffin,B.E. (1979) Coding potential and regulatory signals of the polyomavirus genome. *Nature,* **283**, 445.
96. Muller,W.J., Mueller,C.R., Mes,A.M. and Hassell,J.A. (1983) Polyomavirus origin for DNA replication comprises mutliple genetic elements. *J. Virol.,* **47**, 586.
97. Böhnlein,E. Chowdhury,K. and Gruss,P. (1985) Functional analysis of the regulatory region of polyoma mutant F9-1 DNA. *Nucleic Acids Res.,* **13**, 4789.
98. Mueller,C.R., Mes-Masson,A.-M., Bouvier,M. and Hassell,J.A. (1984) Location of sequences in polyomavirus DNA that are required for early gene expression *in vivo* and *in vitro. Mol. Cell. Biol.,* **4**, 2594.
99. Hearing,P. and Shenk,T. (1983) The adenovirus type 5 E1A transcriptional control region contains a duplicated enhancer element. *Cell,* **33**, 695.
100. Weiher,H. and Botchan,M.R. (1984) An enhancer sequence from bovine papilloma virus DNA consists of two essential regions. *Nucleic Acids Res.,* **12**, 2901.
101. Piette,J. and Yaniv,M. (1987) Two different factors bind to the α-domain of the polyoma virus enhancer, one of which also interacts with the SV40 and c-*fos* enhancers. *EMBO J.,* **6**, 1331.
102. Piette,J., Kryszke,M.-H. and Yaniv,M. (1985) Specific interaction of cellular factors with the B enhancer of polyoma virus. *EMBO J.,* **4**, 2675.
103. Piette,J. and Yaniv,M. (1986) Molecular analysis of the interaction between an enhancer binding factor and its DNA target. *Nucleic Acids Res.,* **14**, 9595.
104. Fujimura,F.K. (1986) Nuclear activity from F9 embryonal carcinoma cells binding specifically to the enhancers of wild-type polyoma virus and PyEC mutant DNA. *Nucleic Acids Res.,* **14**, 2845.
105. Ostapchuk,P., Diffley,J.F.X., Bruder,J.T. Stillman,B., Levine,A.J. and Hearing,P. (1986) Interaction of a nuclear factor with the polyomavirus enhancer region. *Proc. Natl. Acad. Sci. USA,* **83**, 8550.
106. Böhnlein,E. and Gruss,P. (1986) Interaction of distinct nuclear proteins with sequences controlling the expression of polyomavirus early genes. *Mol. Cell. Biol.,* **6**, 1401.
107. de Villiers,J., Schaffner,W., Tyndall,C., Lupton,S. and Kamen,R. (1984) Polyoma virus DNA replication requires an enhancer. *Nature,* **312**, 242.
108. Fujimura,F.K. and Linney,E. (1982) Polyoma mutants that productively infect F9 embryonal carcinoma cells do not rescue wild-type polyoma in F9 cells. *Proc. Natl. Acad. Sci. USA,* **79**, 1479.
109. Wirak,D.O., Chalifour,L.E., Wassarman,P.M., Muller,W.J., Hassell,J.A. and DePamphilis,M.L. (1985) Sequence-dependent DNA replication in preimplantation mouse embryos. *Mol. Cell. Biol.,* **5**, 2924.
110. Campbell,B.A. and Villarreal,L.P. (1985) Host species specificity of polyomavirus DNA replication is not altered by simian virus 40 72-base-pair repeats. *Mol. Cell. Biol.,* **5**, 2924.
111. Jones,K.A., Kadonaga,J.T., Rosenfeld,P.J., Kelly,T.J. and Tjian,T. (1987) A cellular DNA-binding protein that activates eukaryotic transcription and DNA replication. *Cell,* **48**, 79.

112. Campbell,B.A. and Villarreal,L.P. (1986) Lymphoid and other tissue-specific phenotypes of polyomavirus enhancer recombinants: positive and negative combinatorial effects on enhancer specificity and activity. *Mol. Cell. Biol.*, **6**, 2068.

113. Rochford,R., Campbell,B.A. and Villarreal,L.P. (1987) A pancreas specificity results from the combination of polyomavirus and Moloney murine leukemia virus enhancer. *Proc. Natl. Acad. Sci. USA*, **84**, 449.

114. Amati,P. (1985) Polyoma regulatory region: a potential probe for mouse cell differentiation. *Cell*, **43**, 561.

115. Linney,E. and Donerly,S. (1983) DNA fragments from F9 PyEC mutants increase expression of heterologous genes in transfected F9 cells. *Cell*, **35**, 693.

116. Swartzendruber,D.E. and Lehman,J.M. (1975) Neoplastic differentiation: interaction of simian virus 40 and polyoma virus with murine teratocarcinoma cells *in vitro*. *J. Cell. Physiol.*, **85**, 179.

117. Boccara,M. and Kelly,F. (1978) Expression of polyoma virus in heterokaryons between embryonal carcinoma cells and differentiated cells. *Virology*, **90**, 147.

118. Fujimura,F.K., Silbert,P.E., Eckhart,W. and Linney,E. (1981) Polyoma virus infection of retinoic acid-induced differentiated teratocarcinoma cells. *J. Virol.*, **39**, 306.

119. Vasseur,M., Kress,C., Montreau,N. and Blangy,D. (1980) Isolation and characterization of polyoma virus mutants able to develop in embryonal carcinoma cells. *Proc. Natl. Acad. Sci. USA*, **77**, 1068.

120. Katinka,M., Yaniv,M., Vasseur,M. and Blangy,D. (1980) Expression of polyoma early functions in mouse embryonal carcinoma cells depends upon sequence rearrangements in the beginning of the late region. *Cell*, **20**, 393.

121. Katinka,M., Vasseur,M., Montreau,N., Yaniv,M. and Blangy,D. (1981) Polyoma DNA sequences involved in control of viral gene expression in murine embryonal carcinoma cells. *Nature*, **290**, 720.

122. Fujimura,F.K., Deininger,P.L., Friedmann,R. and Linney,E. (1981) Mutation near the polyoma DNA replication origin permits productive infection of F9 embryonal carcinoma cells. *Cell*, **23**, 809.

123. Sekikawa,K. and Levine,A.J. (1981) Isolation and characterization of polyoma host range mutants that replicate in nullipotential embryonal carcinoma cells. *Proc. Natl. Acad. Sci. USA*, **78**, 1100.

124. Maione,R., Passananti,C., De Simone,V., Delli-Bovi,P., Augusti-Tosco,G. and Amati,P. (1985) Selection of mouse neuroblastoma cell-specific polyoma virus mutants with stage differentiative advantages of replication. *EMBO J.*, **4**, 3215.

125. De Simone,V., La Mantia,G., Lania,L. and Amati,P. (1985) Polyomavirus mutation that confers a cell-specific *cis* advantage for viral DNA replication. *Mol. Cell. Biol.*, **5**, 2142.

126. Delli Bovi,P., De Simone,V., Giordano,R. and Amati,P. (1984) Polyomavirus growth and persistence in Friend erythroleukemic cells. *J. Virol.*, **49**, 566.

127. Tanaka,K., Chowdhury,K., Chang,K.S.S., Israel,M. and Ito,Y. (1982) Isolation and characterization of polyoma virus mutants which grow in murine embryonal carcinoma and trophoblast cells. *EMBO J.*, **1**, 1521

128. Melin,F., Pinon,H., Reiss,C., Kress,C., Montreau,N. and Blangy,D. (1985) Common features of polyomavirus mutants selected on PCC4 embryonal carcinoma cells. *EMBO J.*, **4**, 1799.

129. Gorman,C.M., Rigby,P.W.J. and Lane,D.P. (1985) Negative regulation of viral enhancers in undifferentiated embryonic stem cells. *Cell*, **42**, 519.

130. Imperiale,M.J., Hung-Teh,K., Feldman,L.T., Nevins,J.R. and Strickland,S. (1984) Common control of the heat shock gene and early adenovirus genes: evidence for a cellular E1A-like activity. *Mol. Cell. Biol.*, **4**, 867.

131. Borrelli,E., Hen,R. and Chambon,P. (1984) Adenovirus-2 E1A products repress enhancer-induced stimulation of transcription. *Nature*, **312**, 608.

132. Velcich,A. and Ziff,E. (1985) Adenovirus E1A proteins repress transcription from the SV40 early promoter. *Cell*, **40**, 705.

133. Hen,R., Borrelli,E. and Chambon,P. (1985) Repression of the immunoglobulin heavy chain enhancer by the adenovirus-2 E1A products. *Science*, **230**, 1391.

134. Hen,R., Borrelli,E., Fromental,C., Sassone-Corsi,P. and Chambon,P. (1986) A mutated polyoma virus enhancer which is active in undifferentiated embryonal carcinoma cells is not repressed by adenovirus-2 E1A products. *Nature*, **321**, 249.

135. Kovesdi,I., Satake,M., Furukawa,K., Reichel,R., Ito,Y. and Nevins,J.R. (1987) A factor discriminating between the wild-type and a mutant polyomavirus enhancer. *Nature*, **328**, 87.

136. Wasylyk,C., Imler,J.L., Perez-Mutul,J. and Wasylyk,B. (1987) The c-Ha-*ras* oncogene and a tumor promoter activate the polyoma virus enhancer. *Cell*, **48**, 525.

137. Imler,J.L., Schatz,C., Wasylyk,C., Chatton,B. and Wasylyk,B. (1988) A Harvey-*ras* responsive element is also responsive to a tumour-promoter and to serum. *Nature*, **332**, 275.

138. Seif,I., Khoury,G. and Dhar,R. (1979) The genome of human papovavirus BKV. *Cell*, **18**, 963.

139. Yang,R.A.C. and Wu,R. (1979) BK virus DNA: complete nucleotide sequence of a human tumor virus. *Science*, **206**, 456.

140. Rosenthal,N., Kress,M., Gruss,P. and Khoury,G. (1983) BK viral enhancer element and a human cellular homolog. *Science*, **222**, 749.

141. Grinnell,B.W., Berg,D.T. and Walls,J. (1986) Activation of the adenovirus and BK virus late promoters: effects of the BK virus enhancer and *trans*-acting viral early proteins. *Mol. Cell. Biol.*, **6**, 3596.

142. Nowock,J. Borgmeyer,U., Püschel,A.W., Rupp,R.A.W. and Sippel,A.E. (1985) The TGGCA protein binds to the MMTV-LTR, the adenovirus origin of replication, and the BK virus enhancer. *Nucleic Acids Res.*, **13**, 2045.

143. zur Hausen,H. and Glissmann,L. (1979) Lymphotropic papovaviruses isolated from African green monkey and human cells. *Med. Microbiol. Immunol.*, **167**, 137.

144. Takemoto,K.K. Furuno,A., Kato,K. and Yoshiike,K. (1982) Biological and biochemical studies of African green monkey lymphotropic papovavirus. *J. Virol.*, **42**, 502.

145. Pawlita,M., Clad,A. and zur Hausen,H. (1985) Complete DNA sequence of lymphotropic papovavirus: prototype of a new species of the polyomavirus genus. *Virology*, **143**, 196.

146. Mosthaf,L., Pawlita,M. and Gruss,P. (1985) A viral enhancer element specifically active in human haematopoietic cells. *Nature*, **315**, 597.

147. Padgett,B.L., Rogers,C.M. and Walker,D.L. (1977) JC virus, a human polyomavirus associated with progressive multifocal leukoencephalopathy: additional biological characteristics and antigenic relationships. *Infect. Immun.*, **15**, 656.

148. Kenney,S., Natarajan,V., Strike,D., Khoury,G. and Salzman,N.P. (1984) JC virus enhancer-promoter active in human brain cells. *Science*, **226**, 1337.

149. Milner,R.J., Bloom,F.E., Lai,C., Lerner,R.A. and Sutcliffe,J.G. (1984) Brain-specific genes have identifier sequences in their introns. *Proc. Natl. Acad. Sci. USA*, **81**, 713.

150. Miyamura,T., Yoshiike,K. and Takemoto,K.K. (1980) Characterization of JC papovavirus adapted to growth in human embryonic kidney cells. *J. Virol.*, **35**, 498.

151. Miyamura,T., Furuno,A. and Yoshiike,K. (1985) DNA rearrangement in the control region for early transcription in a human polyomavirus JC host range mutant capable of growing in human embryonic kidney cells. *J. Virol.*, **54**, 750.

152. Khalili,K., Rappaport,J. and Khoury,G. (1988) Nuclear factors in human brain cells bind specifically to the JCV regulatory region. *EMBO J.*, **7**, 1205.

153. Lusky,M., Berg,L., Weiher,H. and Botchan,M. (1983) Bovine papilloma virus contains an activator of gene expression at the distal end of the early transcription unit. *Mol. Cell. Biol.*, **3**, 1108.

154. Campo,M.S., Spandidos,D.A., Lang,J. and Wilkie,N.M. (1983) Transcriptional control signals in the genome of bovine papillomavirus type 1. *Nature*, **303**, 77.

155. Spalholz,B.A., Yang,Y.-C. and Howley,P.M. (1985) Transactivation of a bovine papilloma virus transcriptional regulatory element by the E2 gene product. *Cell*, **42**, 183.

156. Haugen,T.H., Cripe,T.P., Ginder,G.D., Karin,M. and Turek,L.P. (1987) *Trans*-activation of an upstream early gene promoter of bovine papilloma virus-1 by a product of the viral E2 gene. *EMBO J.*, **6**, 145.

157. Androphy,E.J., Lowy,D.R. and Schiller,J.T. (1987) Bovine papillomavirus E2 *trans*-activating gene product binds to specific sites in papillomavirus DNA. *Nature*, **325**, 70.

158. Moskaluk,C. and Bastia,D. (1987) The *E2* 'gene' of bovine papillomavirus encodes an enhancer-binding protein. *Proc. Natl. Acad. Sci. USA*, **84**, 1215.

159. Weeks,D.L. and Jones,N.C. (1983) E1A control of gene expression is mediated by sequences 5′ to the transcriptional starts of the early viral genes. *Mol. Cell. Biol.*, **3**, 1222.

160. Sassone-Corsi,P., Hen,R., Borrelli,E., Leff,T. and Chambon,P. (1983) Far upstream sequences are required for efficient transcription from the adenovirus-2 E1A transcription unit. *Nucleic Acids Res.,* **11**, 8735.

161. Hen,R., Borrelli,E., Sassone-Corsi,P. and Chambon,P. (1983) An enhancer element is located 340 base pairs upstream from the adenovirus-2 E1A capsite. *Nucleic Acids Res.,* **11**, 8747.

162. Imperiale,M.J., Feldman,L.T. and Nevins,J.R. (1983) Activation of gene expression by adenovirus and herpesvirus regulatory genes acting *trans* and by a *cis*-acting adenovirus enhancer element. *Cell,* **35**, 127.

163. van Ormondt,H., Maat,J., DeWaard,A. and van der Eb,A.J. (1978) The nucleotide sequence of the transforming Hpal-E fragment of adenovirus type 5 DNA. *Gene,* **4**, 309.

164. Hearing,P. and Shenk,T. (1986) The adenovirus type 5 E1A enhancer contains two functionally distinct domains: one is specific for E1A and the other modulates all early units in *cis*. *Cell,* **45**, 229.

165. Ruley,H.E. and Fried,M. (1983) Sequence repeats in a polyoma virus DNA region important for gene expression. *J. Virol.,* **47**, 233.

166. Goodbourn,S., Zinn,K. and Maniatis,T. (1985) Human β-interferon gene expression is regulated by an inducible enhancer element. *Cell,* **41**, 509.

167. Barrett,P., Clark,L. and Hay,R.T. (1987) A cellular protein binds to a conserved sequence in the adenovirus type 2 enhancer. *Nucleic Acids Res.,* **15**, 2719.

168. Berk,A.J., Lee,F., Harrison,T., Williams,J. and Sharp,P.A. (1979) Pre-early adenovirus 5 gene product regulates synthesis of early viral messenger RNAs. *Cell,* **17**, 935.

169. Jones,N. and Shenk,P. (1979) An adenovirus type 5 early gene function regulates expression of other early viral genes. *Proc. Natl. Acad. Sci. USA,* **76**, 3665.

170. Smith,D.H., Kegler,D.M. and Ziff,E.B. (1985) Vector expression of adenovirus type 5 E1a proteins: evidence for E1a autoregulation. *Mol. Cell. Biol.,* **5**, 2684.

171. Green,M.R., Treisman,R. and Maniatis,T. (1983) Transcriptional activation of cloned human β-globin genes by viral immediate-early gene products. *Cell,* **35**, 137.

172. Gaynor,R.B., Hillman,D. and Berk,A.J. (1984) Adenovirus E1A protein activates transcription of a non-viral gene which is infected or transfected into mammalian cells. *Proc. Natl. Acad. Sci. USA,* **81**, 1193.

173. Svensson,C. and Akusjärvi,G. (1984) Adenovirus 2 early region 1A stimulates expression of both viral and cellular genes. *EMBO J.,* **3**, 789.

174. Kao,H.-T. and Nevins,J.R. (1983) Transcriptional activation and subsequent control of the human heat shock gene during adenovirus infection. *Mol. Cell. Biol.,* **3**, 2058.

175. Stein,R. and Ziff,E.D. (1984) HeLa cell β-tubulin gene transcription is stimulated by adenovirus 5 in parallel with viral early genes by an E1a-dependent mechanism. *Mol. Cell. Biol.,* **4**, 2792.

176. Schrier,P.I., Bernards,R., Vaessen,R.J.M.J., Houweling,A. and van der Eb,A.B. (1983) Expression of class I major histocompatibility antigens switched off by highly oncogenic adenovirus 12 in transformed rat cells. *Nature,* **305**, 771.

177. Borrelli,E., Hen,R., Wasylyk,B. and Chambon,P. (1986) The immunoglobulin heavy chain enhancer is stimulated by the adenovirus type 2 E1A products in mouse fibroblasts. *Proc. Natl. Acad. Sci. USA,* **83**, 2846.

178. Nevins,J.R. (1986) Control of cellular and viral transcription during adenovirus infection. *CRC Crit. Rev. Biochem.,* **19**, 307.

179. Imperiale,M.J. and Nevins,J.R. (1984) Adenovirus 5 E2 transcription unit: an E1A-inducible promoter with an essential element that functions independently of position or orientation. *Mol. Cell. Biol.,* **4**, 875.

180. Murthy,S.C.S., Bhat,G.P. and Thimmappaya,B. (1985) Adenovirus EIIA early promoter: transcriptional control elements and induction by the viral pre-early EIA gene, which appears to be sequence independent. *Proc. Natl. Acad. Sci. USA,* **82**, 2230.

181. Zajchowski,D.A., Boeuf,H. and Kedinger,C. (1985) The adenovirus-2 early EIIa transcription unit possesses two overlapping promoters with different sequence requirements for EIa-dependent stimulation. *EMBO J.,* **4**, 1293.

182. Imperiale,M.J., Hart,R.P. and Nevins,J.R. (1985) An enhancer-like element in the adenovirus E2 promoter contains sequences essential for uninduced and E1A-induced transcription. *Proc. Natl. Acad. Sci. USA,* **82**, 381.

183. Kovesdi,I., Reichel,R. and Nevins,J.R. (1986) Identification of a cellular transcription factor involved in E1A *trans*-activation. *Cell,* **45**, 219.

184. Kovesdi,I., Reichel,R. and Nevins,J.R. (1986) E1A transcription induction: enhanced binding of a factor to upstream promoter sequences. *Science,* **231**, 719.
185. Sivaraman,L., Subramanian,S. and Thimmappaya,B. (1986) Identification of a factor in HeLa cells specific for an upstream transcriptional control sequence of an E1A-inducible adenovirus promoter and its relative abundance in infected and uninfected cells. *Proc. Natl. Acad. Sci. USA,* **83**, 5914.
186. Reichel,R., Kovesdi,I. and Nevins,J.R. (1987) Development control of a promoter-specific factor that is also regulated by the E1A gene product. *Cell,* **48**, 501.
187. La Thangue,N.B. and Rigby,P.W.J. (1987) An adenovirus E1A-like transcription factor is regulated during the differentiation of murine embryonal carcinoma stem cells. *Cell,* **49**, 507.
188. Jalinot,P. and Kedinger,C. (1986) Negative regulatory sequences in the E1a-inducible enhancer of the adenovirus-2 early EIIa promoter. *Nucleic Acids Res.,* **14**, 2651.
189. Lang,J.C., Spandidos,D.A. and Wilkie,N.M. (1984) Transcriptional regulation of a herpes simplex virus immediate early gene is mediated through an enhancer-type sequence. *EMBO J.,* **3**, 389.
190. Preston,C.M. and Tannahill,D. (1984) Effects of orientation and position on the activity of a herpes simplex virus immediate early gene far-upstream region. *Virology,* **137**, 439.
191. Schirm,S., Weber,F., Schaffner,W. and Fleckenstein,B. (1985) A transcription enhancer in the *Herpesvirus samiri* genome. *EMBO J.,* **4**, 2669.
192. Boshart,M., Weber,F., Jahn,G., Dorsch-Häsler,K. Fleckenstein,B. and Schaffner,W. (1985) A very strong enhancer is located upstream of an immediate early gene of human cytomegalovirus. *Cell,* **41**, 521.
193. Dorsch-Häsler,K., Keil,G.M., Weber,F., Jasin,M., Schaffner,W. and Koszinowski,U.H. (1985) A long and complex enhancer activates transcription of the gene coding for the highly abundant immediate early mRNA in murine cytomegalovirus. *Proc. Natl. Acad. Sci. USA,* **82**, 8325.
194. Reisman,D. and Sugden,B. (1986) *Trans* activation of an Epstein – Barr viral transcriptional enhancer by the Epstein – Barr viral nuclear antigen 1. *Mol. Cell. Biol.,* **6**, 3838.
195. McKnight,S.L. Kingsbury,R.C., Spence,A. and Smith,M. (1984) The distal transcription signals of the herpesvirus tk gene share a common hexanucleotide control sequence. *Cell,* **37**, 253.
196. Nelson,J.A. and Groudine,M. (1986) Transcriptional regulation of the human cytomegalovirus major immediate-early gene is associated with induction of DNase I-hypersensitive sites. *Mol. Cell. Biol.,* **6**, 452.
197. Ghazal,P., Lubon,H., Fleckenstein,B. and Hennighausen,L. (1987) Binding of transcription factors and creation of a large nucleoprotein complex on the human cytomegalovirus enhancer. *Proc. Natl. Acad. Sci. USA,* **84**, 3658.
198. Tognoni,A., Cataneo,R., Serfling,E. and Schaffner,W. (1985) A novel expression selection approach allows precise mapping of the hepatitis B enhancer. *Nucleic Acids Res.,* **13**, 7457.
199. Shaul,Y., Rutter,W.J. and Laub,O. (1985) A human hepatitis B enhancer element. *EMBO J.,* **4**, 427.
200. Jameel,S. and Siddiqui,A. (1986) The human hepatitis B virus enhancer requires *trans*-acting cellular factor(s) for activity. *Mol. Cell. Biol.,* **6**, 710.
201. Chang,H.-K., Chou,C.-K. Chang,C. Su,T.-S., Hu,C.-P., Yoshida,M. and Ting,L.-P. (1987) The enhancer sequence of human hepatitis B virus can enhance the activity of its surface gene promoter. *Nucleic Acids Res.,* **15**, 2261.
202. Chandler,V.L., Maler,B.A. and Yamamoto,K.R. (1983) DNA sequences bound specifically by glucocorticoid receptor *in vitro* render a heterologous promoter hormone responsive *in vivo. Cell,* **33**, 489.
203. Ponta,H., Kennedy,N., Skroch,P., Hynes,N.E. and Groner,B. (1985) Hormonal response region in the mouse mammary tumor virus long terminal repeat can be dissociated from the proviral promoter and has enhancer properties. *Proc. Natl. Acad. Sci. USA,* **82**, 1020.
204. Yamamoto,K.R. (1985) Steroid receptor regulated transcription of specific genes and gene networks. *Annu. Rev. Genet.,* **19**, 209.

205. Payvar,F., Defranco,D., Firestone,G.L., Edgar,B., Wrange, Ö., Okret,S., Gustafsson, J.-A. and Yamamoto,K.R. (1983) Sequence-specific binding of glucocorticoid receptor to MTV DNA at sites within and upstream of the transcribed region. *Cell*, **35**, 381.
206. Pfahl,M., McGinnis,D., Hendricks,M., Groner,B. and Hynes,N.E. (1983) Correlation of glucocorticoid receptor binding sites on MMTV proviral DNA with hormone inducible transcription. *Science*, **222**, 1341.
207. Beato,M. (1986) Regulation of gene expression by steroid hormones. In *Oncogenes and Growth Control*. Kahn,P. and Graf,T. (eds), Springler-Verlag, Berlin, p. 219.
208. Godowski,P.J., Rusconi,S., Miesfeld,R. and Yamamoto,K.R. (1987) Glucocorticoid receptor mutants that are constitutive activators of transcriptional enhancement. *Nature*, **325**, 365.
209. Miesfeld,R., Godowski,P.J., Maler,B.A. and Yamamoto,K.R. (1987) Glucocorticoid receptor mutants that define a small region sufficient for enhancer activation. *Science*, **236**, 423.
210. Levinson,B., Khoury,G., Woude,G.V. and Gruss,P. (1982) Activation of SV40 genome by 72-base pair tandem repeats of Moloney sarcoma virus. *Nature*, **295**, 568.
211. Laimins,L.A., Gruss,P., Pozzatti,R. and Khoury,G. (1984) Characterization of enhancer in the long terminal repeat of Moloney murine sarcoma virus. *J. Virol.*, **49**, 183.
212. Defranco,D. and Yamamoto,K.R. (1986) Two different factors act separately or together to specify functionally distinct activities at a single transcriptional enhancer. *Mol. Cell. Biol.*, **6**, 993.
213. Miksicek,R., Heber,A., Schmid,W., Danesch,U., Posseckert,G., Beato,M. and Schutz,G. (1986) Glucocorticoid responsiveness of the transcriptional enhancer of Moloney murine sarcoma virus. *Cell*, **46**, 283.
214. Schulze,F., Böhnlein,E. and Gruss,P. (1985) Mutational analyses of the Moloney murine sarcoma virus enhancer. *DNA*, **4**, 193.
215. Speck,N.A. and Baltimore,D. (1987) Six distinct nuclear factors interact with the 75-base-pair repeat of the Moloney murine leukemia virus enhancer. *Mol. Cell. Biol.*, **7**, 1101.
216. Linney,E., Davis,B., Overhauser,J., Chao,E. and Fan,H. (1984) Non-function of a murine leukemia virus regulatory sequence in F9 embryonal carcinoma cells. *Nature*, **308**, 470.
217. Bösze,Z., Thiesen,H.-J. and Charnay,P. (1986) A transcriptional enhancer with specificity for erythroid cells is located in the long terminal repeat of the Friend murine leukemia virus. *EMBO J.*, **5**, 1615.
218. Haseltine,W.A., Sodroski,J., Rosen,G., Goh,W.C., Dayton,A. and Celander,D. (1986) Transactivator genes of HTLV-I, II, and III. In *Oncogenes and Growth Control*. Kahn,P. and Graf,T. (eds), Springer-Verlag, Berlin, p. 247.
219. Rosen,C.A., Sodroski,J.G. and Haseltine,W.A. (1985) Location of *cis*-acting regulatory sequences in the human T-cell leukemia virus type/long terminal repeat. *Proc. Natl. Acad. Sci. USA*, **82**, 6502.
220. Shimotohno,K., Takano,M., Teruuchi,T. and Miwa,M. (1986) Requirements of multiple copies of a 21-nucleotide sequence in the *U3* regions of human T-cell leukemia virus type I and type II long terminal repeats for *trans*-acting activation of transcription. *Proc. Natl. Acad. Sci. USA*, **83**, 8112.
221. Paskalis,H., Felber,B.K. and Pavlakis,G.N. (1986) *Cis*-acting sequences responsible for the transcriptional activation of human T-cell leukemia virus type I constitute a conditional enhancer. *Proc. Natl. Acad. Sci. USA*, **83**, 6558.
222. Fujisawa,J., Seiki,M., Sato,M. and Yoshida,M. (1986) A transcriptional enhancer sequence of HTLV-I is responsible for *trans*-activation mediated by p40$^\times$ of HTLV-I. *EMBO J.*, **5**, 713.
223. Ohtani,K., Nakamura,M., Saito,S., Noda,T., Ito,Y., Sugamura,K. and Hinuma,Y. (1987) Identification of two distinct elements in the long terminal repeat of HTLV-I responsible for maximum gene expression. *EMBO J.*, **6**, 389.
224. Rosen,C.A. Sodroski,J.G., Kettman,R. and Haseltine,W.A. (1986) Activation of enhancer sequences in type II human T-cell leukemia virus and bovine leukemia virus long terminal repeats by virus-associated *trans*-acting regulatory factors. *J. Virol.*, **57**, 738.
225. Rosen,C.A., Sodroski,J.G. and Haseltine,W.A. (1985) The location of *cis*-acting regulatory sequences in the human T cell lymphotropic virus type III (HTLV-III/LAV) long terminal repeat. *Cell*, **41**, 813.

226. Muesing,M.A., Smith,D.H. and Capon,D.J. (1987) Regulation of mRNA accumulation by a human immunodeficiency virus *trans*-activator protein. *Cell,* **48**, 691.
227. Nabel,G. and Baltimore,D. (1987) An inducible transcription factor activates expression of human immunodeficiency virus in T cells. *Nature,* **326**, 711.
228. Srinivasan,A., Reddy,E.P., Dunn,C.Y. Aaronson,S.A. (1984) Molecular dissection of transcriptional control elements within the long terminal repeat of the retrovirus. *Science,* **223**, 286.
229. Kriegler,M. and Botchan,M. (1983) Enhanced transformation by a simian virus 40 recombinant virus containing a Harvey murine sarcoma virus long terminal repeat. *Mol. Cell. Biol.,* **3**, 325.
230. Luciw,P.A., Bishop,J.M., Varmus,H.E. and Capecchi,M.R. (1983) Location and function of retroviral and SV40 sequences that enhance biochemical transformation after microinjection of DNA. *Cell,* **33**, 705.
231. Laimins,L.A., Tsichlis,P. and Khoury,G. (1984) Multiple enhancer domains in the 3' terminus of the Prague strain of Rous sarcoma virus. *Nucleic Acids Res.,* **12**, 6427.
232. Weber,F. and Schaffner,W. (1985) Enhancer activity correlates with the oncogenic potential of avian retroviruses. *EMBO J.,* **4**, 949.
233. DesGroseillers,L., Rassart,E. and Jolicoeur,P. (1983) Thymotropism of murine leukemia virus is conferred by its long terminal repeat. *Proc. Natl. Acad. Sci. USA,* **80**, 4203.
234. Lenz,J., Celander,D., Crowther,R.L., Patarca,R., Perkins,D.W. and Haseltine,W.A. (1984) Determination of the leukaemogenicity of a murine retrovirus by sequences within the long terminal repeat. *Nature,* **308**, 1984.
235. Banerji,J., Olson,L. and Schaffner,W. (1983) A lymphocyte-specific cellular enhancer is located downstream of the joining region in immunoglobulin heavy chain genes. *Cell,* **33**, 729.
236. Gillies,S.D., Morrison,S.L., Oi,V.T. and Tonegawa,S. (1983) A tissue-specific transcription enhancer element is located in the major intron of a rearranged immuno-globulin heavy chain gene. *Cell,* **33**, 717.
237. Neuberger,M.S. (1983) Expression and regulation of immunoglobulin heavy chain gene transfected into lymphoid cells. *EMBO J.,* **2**, 1373.
238. Mercola,M., Wang,X.-F., Olsen,J. and Calame,K. (1983) Transcriptional enhancer elements in the mouse immunoglobulin heavy chain locus. *Science,* **221**, 663.
239. Queen,C. and Baltimore,D. (1983) Immunoglobulin gene transcription is activated by downstream sequence elements. *Cell,* **33**, 741.
240. Picard,D. and Schaffner,W. (1984) A lymphocyte-specific enhancer in the mouse immunoglobulin *x* gene. *Nature,* **307**, 80.
241. Spandidos,D.A. and Anderson,M.L.M. (1984) A tissue-specific transcription enhancer in the human immunoglobulin λ light chain locus. *FEBS,* **175**, 152.
242. Tonegawa,S. (1983) Somatic generation of antibody diversity. *Nature,* **302**, 575.
243. Mills,F.C., Fisher,L.M., Kuroda,R., Ford,A.M. and Gould,H.J. (1983) DNase I hypersensitive sites in the chromatin of human *x* immunoglobulin heavy-chain genes. *Nature,* **306**, 809.
244. Parslow,T.G. and Graner,D.K. (1983) Structure of a nuclease-sensitive region inside the immunoglobulin kappa gene: evidence for a role in gene regulation. *Nucleic Acids Res.,* **11**, 4775.
245. Grosscheld,R., Weaver,D., Baltimore,D. and Constantini,F. (1984) Introduction of a *x* immunoglobulin gene into the mouse germ line: specific expression in lymphoid cells and synthesis of function antibody. *Cell,* **38**, 647.
246. Bergman,Y., Rice,D., Grosscheld,R. and Baltimore,D. (1984) Two regulatory elements for *x* immunoglobulin gene expression. *Proc. Natl. Acad. Sci. USA,* **81**, 7041.
247. Foster,J., Stafford,J. and Queen,C. (1985) An immunoglobulin promoter displays cell-type specificity independently of the enhancer. *Nature,* **315**, 423.
248. Grosscheld,R. and Baltimore,D. (1985) Cell-type specificity of immunoglobulin gene expression is regulated by at least three DNA sequence elements. *Cell,* **41**, 885.
249. Garcia,J.V., Bich-Thuy,L., Stafford,J. and Queen,C. (1986) Synergism between immunoglobulin enhancers and promoters. *Nature,* **322**, 383.
250. Junker,S. (1982) Phenotype of hybrids between lymphoid cells and rat hepatoma cells. *Exp. Cell Res.,* **139**, 51.

251. Adams,J.M., Harris,A.W., Pinkert,C.A., Corcoran,L.M., Alexander,W.S., Cory,S., Palmiter,R.D. and Brinster,R.L. (1985) The c-*myc* oncogene driven by immunoglobulin enhancers induces lymphoid malignancy in transgenic mice. *Nature*, **318**, 533.

252. Gerlinger,P., LeMeur,M., Irrmann,C., Renard,P., Wasylyk,C. and Wasylyk,B. (1986) B-lymphocyte targeting of gene expression in transgenic mice with the immunoglobulin heavy-chain enhancer. *Nucleic Acids Res.*, **14**, 6565.

253. Wasylyk,C. and Wasylyk,B. (1986) The immunoglobulin heavy-chain B-lymphocyte enhancer efficiently stimulates transcription in non-lymphoid cells. *EMBO J.*, **5**, 553.

254. Kadesch,T., Zervos,P. and Ruezinsky,D. (1986) Functional analysis of the murine IgH enhancer: evidence for negative control of cell-type specificity. *Nucleic Acids Res.*, **14**, 8209.

255. Lenardo,M. , Pierce,J.W. and Baltimore,D. (1987) Protein-binding sites in Ig enhancers determine transcriptional activity and inducibility. *Science*, **236**, 1573.

256. Imler,J.L., Lemaire,C., Wasylyk,C. and Wasylyk,B. (1987) Negative regulation contributes to tissue specificity of the immunoglobulin heavy-chain enhancer. *Mol. Cell. Biol.*, **7**, 2558.

257. Gerster,T., Matthias,P., Thali,M., Jiricny,J. and Schaffner,W. (1987) Cell type-specificity elements of the immunoglobulin heavy chain gene enhancer. *EMBO J.*, **6**, 1323.

258. Wirth,T., Staudt,L. and Baltimore,D. (1987) An octamer oligonucleotide upstream of a TATA motif is sufficient for lymphoid-specific promoter activity. *Nature*, **329**, 174.

259. Mercola,M., Goverman,J., Mirell,C. and Calame,K. (1985) Immunoglobulin heavy-chain enhancer requires one or more tissue-specific factors. *Science*, **227**, 266.

260. Maeda,H., Kitamura,D., Kudo,A., Araki,K. and Watanabe,T. (1986) *Trans*-acting nuclear protein responsible for induction of rearranged human immunoglobulin heavy chain gene. *Cell*, **45**, 25.

261. Schöler,H.R. and Gruss,P. (1985) Cell type-specific transcriptional enhancement *in vitro* requires the presence of *trans*-acting factors. *EMBO J.*, **4**, 3005.

262. Augereau,P. and Chambon,P. (1986) The mouse immunoglobulin heavy-chain enhancer: effect on transcription *in vitro* and binding of proteins present in HeLa and lymphoid B cell extracts. *EMBO J.*, **5**, 1791.

263. Sen,R. Baltimore,D. (1987) *In vitro* transcription of immunoglobulin genes in a B cell extract: effects of enhancer and promoter sequences. *Mol. Cell. Biol.*, in press.

264. Ephrussi,A., Church,G.M., Tonegawa,S. and Gilbert,W. (1985) B lineage-specific interactions of an immunoglobulin enhancer with cellular factors *in vivo. Science*, **227**, 134.

265. Church,G.M., Ephrussi,A., Gilbert,W. and Tonegawa,S. (1985) Cell type specific contacts to immunoglobulin enhancers in nuclei. *Nature*, **313**, 798.

266. Schlokat,U., Bohman,D., Schöler,H. and Gruss,P. (1986) Nuclear factors binding specific sequences within the immunoglobulin enhancer interact differentially with other enhancer elements. *EMBO J.*, **5**, 3251.

267. Singh,H., Sen,R., Baltimore,D. and Sharp,P.A. (1986) A nuclear factor that binds a conserved sequence motif in transcriptional control elements of immunoglobulin genes. *Nature*, **319**, 154.

268. Weinberger,J., Baltimore,D. and Sharp,P.A. (1986) Distinct factors bind to apparently homologous sequences in the immunoglobulin heavy-chain enhancer. *Nature*, **322**, 846.

269. Staudt,L.M. Singh,H., Sen,R., Wirth,T., Sharp,P.A. and Baltimore,D. (1986) A lymphoid-specific protein binding to the octamer motif of immunoglobulin genes. *Nature*, **323**, 640.

270. Landolfi,N.F., Capra,J.D. and Tucker,P.A. (1986) Interaction of cell-type-specific nuclear proteins with immunoglobulin V_H promoter region sequences. *Nature*, **323**, 548.

271. Peterson,G.L., Orth,K. and Calame,K.L. (1986) Binding *in vitro* of multiple cellular proteins to immunoglobulin heavy-chain enhancer DNA. *Mol. Cell. Biol.*, **6**, 4168.

272. Hromas,R. and Van Ness,B. (1986) Nuclear factors bind to regulatory regions of the mouse kappa immunoglobulin gene. *Nucleic Acids Res.*, **14**, 4837.

273. Mocikat,R., Falkner,F.G. and Zachau,H.G. (1986) Upstream regulatory sequences of immunoglobulin genes are recognized by nuclear proteins which also bind to other gene regions. *Nucleic Acids Res.*, **14**, 8829.

274. Sive,H.L. and Roeder,R.G. (1986) Interaction of a common factor with conserved promoter and enhancer sequences in histone H2B, immunoglobulin, and U2 small nuclear RNA (snRNA) genes. *Proc. Natl. Acad. Sci. USA*, **83**, 6382.

275. Pruijn,G.J.M., van Driel,W. and van der Vliet,P.C. (1986) Nuclear factor III, a novel sequence-specific DNA-binding protein from HeLa cells stimulating adenovirus DNA replication. *Nature*, **322**, 656.

276. Peterson,C.L. and Calame,K.L. (1987) Complex protein binding within the mouse immunoglobulin heavy-chain enhancer. *Mol. Cell. Biol.*, **7**, 4194.

277. Sen,R. and Baltimore,D. (1986) Inducibility of ϰ immunoglobulin enhancer-binding protein NF-ϰB by a posttranslational mechanism. *Cell*, **47**, 921.

278. Wall,R. Briskin,M., Carter,C., Goven,H., Taylor,A. and Kincade,P. (1986) A labile inhibitor blocks immunoglobulin ϰ-light-chain-gene transcription in a pre-B leukemic cell line. *Proc. Natl. Acad. Sci. USA*, **83**, 295.

279. Gerster,T., Picard,D. and Schaffner,W. (1986) During B-cell differentiation enhancer activity and transcription rate of immunoglobulin heavy chain genes are high before mRNA accumulation. *Cell*, **45**, 45.

280. Atchison,M.L. and Perry,R.P. (1987) The role of the ϰ enhancer and its binding factor NF-ϰB in the developmental regulation of ϰ gene transcription. *Cell*, **48**, 121.

281. Wang,X.-F. and Calame,K. (1986) SV40 enhancer-binding factors are required at the establishment but not the maintenance step of enhancer-dependent transcriptional activation. *Cell*, **47**, 241.

282. Wabl,M.R. and Burrows,P.D. (1984) Expression of immunoglobulin heavy chain at a high level in the absence of a proposed immunoglobulin enhancer element in *cis*. *Proc. Natl. Acad. Sci. USA*, **81**, 2452.

283. Pierce,J.W., Lenardo,M. and Baltimore,D. (1988) Oligonucleotide that binds nuclear factor NF-ϰB acts as a lymphoid-specific and inducible enhancer element. *Proc. Natl. Acad. Sci. USA*, **85**, 1482.

284. Walker,M.D., Edlund,T., Boulet,A.M. and Rutter,W.J. (1983) Cell-specific expression controlled by the 5'-flanking region of insulin and chymotrypsin genes. *Nature*, **306**, 557.

285. Swift,G.H., Hammer,R.E., MacDonald,R.J. and Brinster,R.L. (1984) Tissue-specific expression of the rat pancreatic elastase I gene in transgenic mice. *Cell*, **38**, 639.

286. Boulet,A.M., Erwin,C.R. and Rutter,W.J. (1986) Cell-specific enhancers in the rat exocrine pancreas. *Proc. Natl. Acad. Sci. USA*, **83**, 3599.

287. Ornitz,D.M., Palmiter,R.D., Hammer,R.E., Brinster,R.L., Swift,G.H. and MacDonald,R.J. (1985) Specific expression of an elastase-human growth hormone fusion gene in pancreatic acinar cells of transgenic mice. *Nature*, **313**, 600.

288. Hanahan,D. (1985) Heritable formation of pancreatic β-cell tumours in transgenic mice expressing recombinant insulin/simian virus 40 oncogenes. *Nature*, **315**, 115.

289. Edlund,T., Walker,M.D., Barr,P.J. and Rutter,W.J. (1985) Cell-specific expression of the rat insulin gene: evidence for role of two distinct 5' flanking elements. *Science*, **230**, 912.

290. Hammer,R.E., Swift,G.H., Ornitz,D.M., Quaife,C.J., Palmiter,R.D., Brinstel,R.L. and MacDonald,R.J. (1987) The rat elastase I regulatory element is an enhancer that directs correct cell specificity and developmental onset of expression in transgenic mice. *Mol. Cell. Biol.*, **7**, 2956.

291. Nir,U., Walker,M.D. and Rutter,W.J. (1986) Regulation of rat insulin 1 gene expression: evidence for negative regulation in nonpancreatic cells. *Proc. Natl. Acad. Sci. USA*, **83**, 3180.

292. Laimins,L., Homgren-König,M. and Khoury,G. (1986) Transcriptional 'silencer' element in rat repetitive sequences associated with the rat insulin 1 gene locus. *Proc. Natl. Acad. Sci. USA*, **83**, 3151.

293. Ohlsson,H. and Edlund,T. (1986) Sequence-specific interactions of nuclear factors with the insulin gene enhancer. *Cell*, **45**, 35.

294. Imbra,R.J. and Karin,M. (1987) Metallothionein gene expression is regulated by serum factors and activators of protein kinase C. *Mol. Cell. Biol.*, **7**, 1358.

295. Karin,M., Haslinger,A., Holtgreve,H., Richards,R.I., Krauter,P., Westphal,H.M. and Beato,M. (1984) Characterization of DNA sequences through which cadmium and glucocorticoid hormones induce human metallothionein-II$_A$ gene. *Nature*, **308**, 513.

296. Haslinger,A. and Karin,M. (1985) Upstream promoter element of the human metallothionein-II$_A$ gene can act like an enhancer element. *Proc. Natl. Acad. Sci. USA*, **82**, 8572.

297. Serfling,E., Lübbe,A., Dorsch-Häsler,K. and Schaffner,W. (1985) Metal dependent SV40 viruses containing inducible enhancers from the upstream region of metallothionein genes. *EMBO J.*, **4**, 3851.

298. Seguin,C. and Hamer,D.H. (1987) Regulation *in vitro* of metallothionein gene binding factors. *Science*, **235**, 1383.

299. Andersen,R.D., Taplitz,S.T., Wong,S., Bristol,G., Larkin,B. and Herschman,H.R. (1987) Metal-dependent binding of a factor *in vivo* to the metal-responsive elements of the metallothionein 1 gene promoter. *Mol. Cell. Biol.*, **7**, 3574.

300. Lengyel,P. (1982) Biochemistry of interferons and their actions. *Annu. Rev. Biochem.*, **51**, 251.

301. Enoch,T., Zinn,K. and Maniatis,T. (1986) Activation of the human β-interferon gene requires an interferon-inducible factor. *Mol. Cell. Biol.*, **6**, 801.

302. Ragg,H. and Weissmann,C. (1983) Not more than 117 base pairs of 5' flanking sequence are required for inducible expression of a human IFN-α gene. *Nature*, **303**, 439.

303. Maroteaux,L., Kahana,C., Mory,Y., Revel,M. and Howley,P. (1983) Sequences involved in the regulated expression of the human interferon-β1 gene in recombinant SV40 DNA vectors replicating in monkey cells. *EMBO J.*, **2**, 325.

304. Zinn,K., DiMaio,D. and Maniatis,T. (1983) Identification of two distinct regulatory regions adjacent to the human β-interferon gene. *Cell*, **34**, 865.

305. Ohno,S. and Taniguchi,T. (1983) The 5' flanking sequence of human interferon-1 gene is responsible for viral induction of transcription. *Nucleic Acids Res.*, **11**, 5403.

306. Fujita,T., Ohno,S., Yasumitsu,H. and Taniguchi,T. (1985) Delimination and properties of DNA sequences required for the regulated expression of human interferon-β gene. *Cell*, **41**, 489.

307. Fujita,T., Shibuya,H., Hotta,H., Yamanishi,K. and Taniguchi,T. (1987) Interferon-β regulation: tandemly repeated sequences of a synthetic 6 bp oligomer function as a virus-inducible enhancer. *Cell*, **49**, 357.

308. Ryals,J., Dierks,P., Ragg,H. and Weissmann,C. (1985) A 46-nucleotide promoter segment for an IFN-α gene renders an unrelated promoter inducible by virus. *Cell*, **41**, 497.

309. Goodbourn,S., Burstein,H. and Maniatis,T. (1986) The human β-interferon gene enhancer is under negative control. *Cell*, **45**, 601.

310. Zinn,K. and Maniatis,T. (1986) Detection of factors that interact with the human β-interferon regulatory region *in vivo* by DNase I footprinting. *Cell*, **45**, 611.

311. Goodbourn,S. and Maniatis,T. (1988) Overlapping positive and negative regulatory domains of the human β-interferon gene. *Proc. Natl. Acad. Sci. USA*, **85**, 1447.

312. Franza,B.R.,Jr, Rauscher,F.J.,III, Josephs,S.F. and Curran,T. (1988) The *fos* complex and *fos*-related antigens recognize sequence elements that contain AP-1 binding sites. *Science*, **239**, 1150.

313. Müller,R., Verma,I.M. and Adamson,E.D. (1983) Expression of c-*onc* genes: c-*fos* transcripts accumulate to high levels during development of mouse placenta, yolk sac and amnion. *EMBO J.*, **2**, 679.

314. Adamson,E.D., Meek,J. and Edwards,S.A. (1985) Product of the cellular oncogene, c-*fos*, observed in mouse and human tissues using an antibody to a synthetic peptide. *EMBO J.*, **4**, 941.

315. Treisman,R. (1985) Transient accumulation of c-*fos* RNA following serum stimulation requires a conserved 5' element and c-*fos* 3' sequences. *Cell*, **42**, 889.

316. Deschamps,J., Meijilink,F. and Verma,I.M. (1985) Identification of a transcriptional enhancer element upstream from the proto-oncogene *fos*. *Science*, **230**, 1174.

317. Renz,M., Neuberg,M., Kurz,C., Bravo,R. and Müller,R. (1985) Regulation of c-*fos* transcription in mouse fibroblasts: identification of DNase I-hypersensitive sites and regulatory upstream sequences. *EMBO J.*, **4**, 3711.

318. Sassone-Corsi,P. and Verma,I.M. (1987) Modulation of c-*fos* gene transcription by negative and positive cellular factors. *Nature*, **326**, 507.

319. Treisman,R. (1986) Identification of a protein-binding site that mediates transcriptional response of the c-*fos* gene to serum factors. *Cell*, **46**, 567.

320. Prywes,R. and Roeder,R.G. (1986) Inducible binding of a factor to the c-*fos* enhancer. *Cell*, **47**, 777.

321. Hayes,T.E. Kitchen,A.M., Cochran,B.H. (1987) Inducible binding of a factor to the c-*fos* regulatory region. *Proc. Natl. Acad. Sci. USA*, **84**, 1272.

322. Greenberg,M.E., Siegfried,Z. and Ziff,E.B. (1987) Mutation of the c-*fos* gene dyad symmetry element inhibits serum inducibility of transcription *in vivo* and the nuclear regulatory factor binding *in vitro*. *Mol. Cell. Biol.*, **7**, 1217.

323. Prywes,R. and Roeder,R.G. (1987) Purification of the c-*fos* enhancer-binding protein. *Mol. Cell. Biol.*, **7**, 3482.

324. Treisman,R. (1987) Identification and purification of a polypeptide that binds to the c-*fos* serum response element. *EMBO J.*, **6**, 2711.

325. Gilman,M.Z., Wilson,R.N. and Weinberg,R.A. (1986) Multiple protein-binding sites in the 5' flanking region regulate c-*fos* expression. *Mol. Cell. Biol.*, **6**, 4305

326. Tilghman,S.M. and Belayew,A. (1982) Transcriptional control of the murine albumin/α-fetoprotein locus during development. *Proc. Natl. Acad. Sci. USA*, **79**, 5254.

327. Scott,R.W. Vogt,T.F., Croke,M.E. and Tilghman,S.M. (1984) Tissue-specific activation of a cloned α-fetoprotein gene during differentiation of a transfected embryonal carcinoma cell line. *Nature*, **310**, 562.

328. Godbout,R., Ingram,R. and Tilghman,S.M. (1986) Multiple regulatory elements in the intergenic region between the α-fetoprotein and albumin genes. *Mol. Cell. Biol.*, **6**, 477.

329. Widen,S.G. and Papaconstantinou,J. (1986) Liver-specific expression of the mouse α-fetoprotein gene is mediated by *cis*-acting DNA elements. *Proc. Natl. Acad. Sci. USA*, **83**, 8196.

330. Hammer,R.E., Krumlauf,R., Camper,S.A., Brinster,R.L. and Tilghman,S.M. (1987) Diversity of alpha-fetoprotein gene expression in mice is generated by a combination of separate enhancer elements. *Science*, **235**, 53.

331. Godbout,R., Ingram,R.S. and Tilghman,S.M. (1988) Fine-structure mapping of the three mouse α-fetoprotein gene enhancers. *Mol. Cell. Biol.*, **8**, 1169.

332. Kioussis,D., Eiferman,F., van de Rijn,P., Gorin,M.B., Ingram,R.S. and Tilghman,S.M. (1981) The evolution of α-fetoprotein and albumin. *J. Biol. Chem.*, **256**, 1960.

333. Pinkert,C.A., Ornitz,D.M., Brinster,R.L. and Palmiter,R.D. (1987) An albumin enhancer located 10 kb upstream functions along with its promoter to diet efficient, liver-specific expression in transgenic mice. *Genes and Development*, **1**, 268.

334. Costa,R.H., Lai,E. and Darnell,J.E.,Jr (1986) Transcriptional control of the mouse prealbumin (transthyretin) gene: both promoter sequences and a distinct enhancer are cell specific. *Mol. Cell. Biol.*, **6**, 4697.

335. Costa,R.H., Lai,E., Grayson,D.R. and Darnell,J.E.,Jr (1988) The cell-specific enhancer of the mouse transthyretin (prealbumin) gene binds a common factor at one site and a liver-specific factor(s) at two other sites. *Mol. Cell. Biol.*, **8**, 81.

336. Grayson,D.R., Costa,R.H., Xanthopoulos,K.G. and Darnell,J.E.,Jr (1988) A cell-specific enhancer of the mouse α1-antitrypsin gene has multiple functional regions and corresponding protein-binding sites. *Mol. Cell. Biol.*, **8** 1055.

337. De Simone,V., Ciliberto,G., Hardon,E., Paonessa,G., Palla,F., Lundberg,L. and Cortese,R. (1987) *Cis*- and *trans*-acting elements for the cell-specific expression of the human α1-antitrypsin gene. *EMBO J.*, **6**, 2759.

338. Grayson,D.R., Costa,R.H., Xanthopoulos,K.G. and Darnell,J.E. (1988) One factor recognizes the liver-specific enhancers in α_1-antitrypsin and transthyretin genes. *Science*, **239**, 786.

339. Hood,L., Steinmetz,M. and Mallisen,B. (1983) Genes of the MHC of mouse. *Annu. Rev. Immunol.*, **1**, 529.

340. Rosa,F., Le Bouteiller,D., Abadie,A., Mishal,Z., Lemonnier,F.A., Bourrel,D., Lamotte,M., Kali,J., Jordan,B.R. and Fellous,M. (1983) HLA class I genes integrated into murine cells are inducible by interferon. *Eur. J. Immunol.*, **13**, 495.

341. Morello,D., Daniel,F., Baldacci,P., Cayre,Y., Gachelin,G. and Kourilsky,P. (1982) Absence of significant H-2 and β_2-microglobulin mRNAs expression by mouse embryonal carcinoma cells. *Nature*, **296**, 260.

342. Rosenthal,A., Wright,S., Cedar,H., Flavell,R. and Grosveld,F. (1984) Regulated expression of an introduced MHC *H-2K^{bm1}* gene in murine embryonal carcinoma cells. *Nature*, **310**, 415.

343. Kimura,A., Israel,A., Le Bail,O. and Kourilsky,P. (1986) Detailed analysis of the mouse H-2K^b promoter: enhancer-like sequences and their role in the regulation of class I gene expression. *Cell*, **44**, 261.

344. Miyazaki,J.-I., Appella,E. and Ozato,K. (1986) Negative regulation of the major histocompatibility class I gene in undifferentiated embryonal carcinoma cells. *Proc. Natl. Acad. Sci. USA*, **83**, 9537.

345. Vogel,J., Kress,M., Khoury,G. and Jay,G. (1986) A transcriptional enhancer and an interferon-responsive sequence in major histocompatibility complex class I genes. *Mol. Cell. Biol.*, **6**, 3550.

346. Israel,A., Kimura,A., Fournier,A., Fellous,M. and Kourilsky,P. (1986) Interferon response sequence potentiates activity of an enhancer in the promoter region of a mouse *H-2* gene. *Nature*, **322**, 743.

347. Baldwin,A.S.,Jr and Sharp,P.A. (1987) Binding of a nuclear factor to a regulatory sequence in the promoter of the mouse *H-2K^b* class I major histocompatibility gene. *Mol. Cell. Biol.*, **7**, 305.

348. Israel,A., Kimura,A., Kieran,M., Yano,O., Kanellopoulos,J., Le Bail,O. and Kourilsky,P. (1987) A common positive *trans*-acting factor binds to enhancer sequences in the promoters of mouse *H-2* and β_2-microglobulin genes. *Proc. Natl. Acad. Sci. USA*, **84**, 2653.

349. Baldwin,A.S.,Jr and Sharp,P.A. (1988) Two transcriptional factors, NF-κB and H2TF1, interact with a single regulatory sequence in the class I major histocompatibility complex promoter. *Proc. Natl. Acad. Sci. USA*, **85**, 723.

350. Yano,O., Kanellopoulos,J., Kieran,M., Le Bail,O., Israel,A. and Kourilsky,P. (1987) Purification of KBF1, a common factor binding to both H-2 and β_2-microglobulin enhancers. *EMBO J.*, **6**, 3317.

351. Singh,H., LeBowitz,J.H., Baldwin,A.S.,Jr and Sharpe,P.A. (1988) Molecular cloning of an enhancer binding protein: isolation by screening of an expression library with a recognition site DNA. *Cell*, **52**, 415.

352. Gillies,S.D., Folsom,V. and Tonegawa,S. (1984) Cell type-specific enhancer element associated with a mouse MHC gene, E_β. *Nature*, **310**, 594.

353. Sullivan,K.E. and Peterlin,B.M. (1987) Transcriptional enhancers in the HLA-DQ subregion. *Mol. Cell. Biol.*, **7**, 3315.

354. Nelson,C., Crenshaw,E.B., III, Franco,R., Lira,S.A., Albert,V.R., Evans,R.M. and Rosenfeld,M.G. (1986) Discrete *cis*-active genomic sequences dictate the pituitary cell type-specific expression of rat prolactin and growth hormone genes. *Nature*, **322**, 557.

355. Elsholtz,H.P., Mangalam,H.J., Potter,E., Albert,V.R., Supowit,S., Evans,R.M. and Rosenfeld,M.G. (1986) Two different *cis*-active elements transfer the transcriptional effects of both EGF and phorbol esters. *Science*, **234**, 1552.

356. Larsen,P.R., Harney,J.W. and Moore,D.D. (1986) Repression mediates cell-type-specific expression of the rat growth hormone gene. *Proc. Natl. Acad. Sci. USA*, **83**, 8283.

357. Lefevre,C., Imagawa,M., Dana,S., Grindlay,J., Bodner,M. and Karin,M. (1987) Tissue-specific expression of the human growth hormone gene is conferred in part by the binding of a specific *trans*-acting factor. *EMBO J.*, **6**, 971.

358. Nelson,C., Albert,V.R., Elsholtz,H.P., Lu,L.I.-W. and Rosenfeld,G. (1988) Activation of cell-specific expression of rat growth hormone and prolactin genes by a common transcription factor. *Science*, **239**, 1400.

359. Comb,M., Birnberg,N.C., Seasholtz,A., Herbert,E. and Goodman,H.M. (1986) A cyclic AMP- and phorbol ester-inducible DNA element. *Nature*, **323**, 353.

360. Montminy,M.R. and Bilezikjian,L.M. (1987) Binding of a nuclear protein to the cyclic-AMP response element of the somatostatin gene. *Nature*, **328**, 175.

361. Delegeane,A.M., Ferland,L.H. and Mellon,P.L. (1987) Tissue-specific enhancer of the human glycoprotein hormone α-subunit gene: dependence on cyclic AMP-inducible elements. *Mol. Cell. Biol.*, **7**, 3994.

362. Prochownik,E.V. (1985) Relationship between an enhancer element in the human antithrombin III gene and an immunoglobulin light-chain gene enhancer. *Nature*, **316**, 845.

363. Theisen,M., Stief,A. and Sippel,A.E. (1986) The lysozyme enhancer: cell-specific activation of the chicken lysozyme gene by a far-upstream element. *EMBO J.*, **5**, 719.

364. Okazaki,K., Yasuda,K., Kondoh,H. and Okada,T.S. (1985) DNA sequences responsible for tissue-specific expression of a chicken α-crystallin gene in mouse lens cells. *EMBO J.*, **4**, 2589.

365. Hayashi,S., Goto,K., Okada,T.S. and Kondoh,H. (1987) Lens-specific enhancer in the third intron regulates expression of the chicken δ1-crystallin gene. *Genes and Development*, **1**, 818.
366. Jones,P.B.C., Durrin,L.K., Galeazzi,D.R. and Whitlock,J.P.,Jr (1986) Control of cytochrome P₁-450 gene expression: analysis of a dioxin-responsive enhancer system. *Proc. Natl. Acad. Sci. USA*, **83**, 2802.
367. Kawamoto,T., Makino,K., Niwa,H., Sugiyama,H., Kimura,S., Amemura,M., Nakata,A. and Kakunaga,T. (1988) Identification of the human β-actin enhancer and its binding factor. *Mol. Cell. Biol.*, **8**, 267.
368. Rossi,P. and De Crombrugghe,B. (1987) Identification of a cell-specific transcriptional enhancer in the first intron of the mouse α₂ (type I) collagen gene. *Proc. Natl. Acad. Sci. USA*, **84**, 5590.
369. Jaynes,J.B., Johnson,J.E., Buskin,J.N., Gartside,C.L. and Hauschka,S.D. (1988) The muscle creatine kinase gene is regulated by multiple upstream elements, including a muscle-specific enhancer. *Mol. Cell. Biol.*, **8**, 62.
370. Choi,O.-R. and Engel,J.D. (1986) A 3' enhancer is required for temporal and tissue-specific transcriptional activation of the chicken adult β-globin gene. *Nature*, **323**, 731.
371. Hesse,J.E., Nickol,J.M., Lieber,M.R. and Felsenfeld,G. (1986) Regulated gene expression in transfected primary chicken erythrocytes. *Proc. Natl. Acad. Sci. USA*, **83**, 4312.
372. Trudel,M. and Constantini,F. (1987) A 3' enhancer contributes to the stage-specific expression of the human β-globin gene. *Genes and Development*, **1**, 954.
373. Antoniou,M., deBoer,E., Habets,G. and Grosveld,F. (1988) The human β-globin gene contains multiple regulatory regions: identification of one promoter and two downstream enhancers. *EMBO J.*, **7**, 377.
374. Bodine,D.M. and Ley,T.J. (1987) An enhancer element lies 3' to the human ᴬγ globin gene. *EMBO J.*, **6**, 2997.
375. Kretsovali,A., Müller,M.M., Weber,F., Marcaud,L., Farache,G., Schreiber,E., Schaffner,W. and Scherrer,K. (1987) Transcriptional enhancer located between the avian adult β-globin and the embryonic ε-globin genes. *Gene*, **58**, 167.
376. Trainor,C.D., Stamler,S.J. and Engel,J.D. (1987) Erythroid-specific transcription of the histone H5 gene is directed by a 3' enhancer. *Nature*, **328**, 827.
377. Garabedian,M.J., Shepherd,B.M. and Wensink,P.C. (1986) A tissue-specific transcription enhancer from the Drosophila yolk protein 1 gene. *Cell*, **45**, 859.
378. Shermoen,A.W., Jongens,J., Barnett,S.W., Flynn,K. and Bechendorf,S.K. (1987) Developmental regulation by an enhancer from the Sgs-4 gene of Drosophila. *EMBO J.*, **6**, 207.
379. Hiromi,Y., Kuroiwa,A. and Gehring,W.J. (1985) Control elements of the Drosophila segmentation gene fushi tarazu. *Cell*, **43**, 603.
380. Bienz,M. and Pelham,H.R.B. (1986) Heat shock regulatory elements function as an inducible enhancer in the Xenopus hsp70 gene and when linked to a heterologous promoter. *Cell*, **45**, 753.
381. Guarente,L. and Hoar,E. (1984) Upstream activation sites of the CYC1 gene of Saccharomyces cerevisiae are active when inverted but not when placed downstream of the 'TATA box'. *Proc. Natl. Acad. Sci. USA*, **81**, 7860.
382. Struhl,K. (1984) Genetic properties and chromatin structure of the yeast gal regulatory element: an enhancer-like sequence. *Proc. Natl. Acad. Sci. USA*, **81**, 7865.
383. Timko,M.P., Kausch,A.P., Castresana,C., Fassler,J., Herrera-Estrella,L. Van den Broeck,J.G., Van Montagu,M., Schell,J. and Cashmore,A.R. (1985) Light regulation of plant gene expression by an upstream enhancer-like element. *Nature*, **318**, 579.
384. Simpson,J., Schell,J., Van Montagu,M. and Herrara-Estrella,L. (1986) Light-inducible and tissue-specific pea Ihcp gene expression involves an upstream element combining enhancer- and silencer-like properties. *Nature*, **323**, 551.
385. Conrad,S.E. and Botchan,M.R. (1982) Isolation and characterization of human DNA fragments with nucleotide homologies with the simian virus 40 regulatory region. *Mol. Cell. Biol.*, **2**, 949.
386. Fried,M., Griffiths,M., Davies,B., Bjursell,G., La Mantia,G. and Lania,L. (1983) Isolation of cellular DNA sequences that allow expression of adjacent genes. *Proc. Natl. Acad. Sci. USA*, **80**, 2117.

387. von Hoyningen-Huene,V., Norbury,C., Griffiths,M. and Fried,M. (1986) Gene activation properties of a mouse DNA sequence isolated by expression selection. *Nucleic Acids Res.,* **14**, 5615.
388. Williams,T.J. and Fried,M. (1986) The MES-1 murine enhancer is closely associated with the heterogeneous 5' ends of two divergent transcription units. *Mol. Cell. Biol.,* **6**, 4558.
389. Hamada,H. (1986) Random isolation of gene activator elements from the human genome. *Mol. Cell. Biol.,* **6**, 4185.
390. Parslow,T.G., Jones,S.D., Bond,B. and Yamamoto,K.R. (1987) The immunoglobulin octanucleotide: independent activity and selective interaction with enhancers. *Science,* **235**, 1499.
391. Serfling,E., Jasin,M. and Schaffner,W. (1985) Enhancers and eukaryotic gene transcription. *Trends Genet.,* **1**, 224.
392. Voss,S.D., Schlokat,U. and Gruss,P. (1986) The role of enhancers in the regulation of cell-type-specific transcriptional control. *Trends Biochem. Sci.,* **11**, 287.
393. Maniatis,T., Goodbourn,S. and Fischer,J.A. (1987) Regulation of inducible and tissue-specific gene expression. *Science,* **236**, 1237.
394. Schirm,S., Jiricny,J. and Schaffner,W. (1987) The SV40 enhancer can be dissected into multiple segments, each with a different cell type specificity. *Genes and Development,* **1**, 65.
395. Ondek,B., Shepard,A. and Herr,W. (1987) Discrete elements within the SV40 enhancer region display different cell-specific activities. *EMBO J.,* **6**, 1017.
396. Toohey,M.G., Morley,K.L. and Peterson,D.O. (1986) Multiple hormone-inducible enhancers as mediators of differential transcription. *Mol. Cell. Biol.,* **6**, 4526.
397. Ryoji,M. and Worcel,A. (1984) Chromatin assembly in *Xenopus* oocytes: *in vivo* studies. *Cell,* **37**, 21.
398. Villeponteau,B., Lundrell,M. and Martinson,H. (1984) Torsional stress promotes the DNase I sensitivity of active genes. *Cell,* **39**, 469.
399. Yang,L., Rowe,T.C., Nelson,E.M. and Liu,L.F. (1985) *In vivo* mapping of topoisomerase II-specific cleavage sites on SV40 chromatin. *Cell,* **41**, 127.
400. Plon,S.E. and Wang,J.C. (1986) Transcription of the human β-globin gene is stimulated by an SV40 enhancer to which is physically linked but topologically uncoupled. *Cell,* **45**, 575.
401. Hochschild,A., Irwin,N. and Ptashne,M. (1983) Repressor structure and the mechanism of positive control. *Cell,* **32**, 319.
402. Brent,R. and Ptashne,M. (1985) A eukaryotic transcription activator bearing the DNA specificity of a prokaryotic repressor. *Cell,* **43**, 729.
403. Herbomel,P., Saragosti,S., Blangy,D. and Yaniv,M. (1981) Fine structure of the origin-proximal DNase I-hypersensitive region in wild-type and EC mutant polyoma. *Cell,* **25**, 651.
404. Abulafia,R., Ben-Ze'ev,A., Hay,N. and Aloni,Y. (1984) Control of late simian virus 40 transcription by the attenuation mechanism and transcriptionally active ternary complexes are associated with the nuclear matrix *J. Mol. Biol.,* **172**, 467.
405. Robinson,S.I., Small,D., Idzerda,R., McKnight,G.S. and Vogelstein,B. (1983) The association of transcriptionally active genes with the nuclear matrix of the chicken oviduct. *Nucleic Acids Res.,* **11**, 5113.
406. Jost,J. and Seldran,M. (1984) Association of transcriptionally active vitellogenin II gene with the nuclear matrix of chicken liver. *EMBO J.,* **3**, 2005.
407. Mirkovitch,J., Mirault,M.-E. and Laemmli,U.K. (1984) Organization of the higher-order chromatin loop: specific DNA attachment sites on nuclear scaffold. *Cell,* **39**, 223.
408. Jackson,D.A. and Cook,P.R. (1985) Transcription occurs at a nucleoskeleton. *EMBO J.,* **4**, 919.
409. Gasser,S.M. and Laemmli,U. (1986) Cohabitation of scaffold binding regions with upstream/enhancer elements of three developmentally regulated genes of D.melanogaster. *Cell,* **46**, 521.
410. Cockerill,P.N. and Garrard,W.T. (1986) Chromosomal loop anchorage of the kappa immunoglobulin gene occurs next to the enhancer in a region containing topoisomerase II sites. *Cell,* **44**, 273.
411. Sassone-Corsi,P. and Borrelli,E. (1986) Transcriptional regulation by *trans*-acting factors. *Trends Genet.,* **2**, 215.

412. Ptashne,M. (1986) Gene regulation by proteins acting nearby and at a distance. *Nature,* **322**, 697.
413. Hochschild,A. and Ptashne,M. (1986) Cooperative binding of repressors to sites separated by integral turns of the DNA helix. *Cell,* **44**, 681.
414. Griffith,J., Hochschild,A. and Ptashne,M. (1986) DNA loops induced by cooperative binding of λ repressor. *Nature,* **322**, 750.
415. Majumdar,A. and Adhya,S. (1984) Demonstration of two operator elements in *gal: in vitro* repressor binding studies. *Proc. Natl. Acad. Sci. USA,* **81**, 6100.
416. Martin,K., Huo,L. and Schleif,R.F. (1986) The DNA loop model for *ara* repression: AraC protein occupies the proposed loop sites *in vivo* and repression-negative mutations lie in these same sites. *Proc. Natl. Acad. Sci. USA,* **83**, 3654.
417. Jones,K.A. Yamamoto,K.R. and Tjian,R. (1985) Two distinct transcription factors bind to the HSV thymidine kinase promoter *in vitro. Cell,* **42**, 559.
418. Graves,B.J., Johnson,P.F. and McKnight,S.L. (1986) Homologous recognition of a promoter domain common to the MSV LTR and the HSV *tk* gene. *Cell,* **44**, 565.
419. Dorn,A., Bollekens,J., Staub,A., Benoist,C. and Mathis,D. (1987) A multiplicity of CCAAT box-binding proteins. *Cell,* **50**, 863.
420. Raymondjean,M., Cereghini,S. and Yaniv,M. (1988) Several distinct 'CCAAT' box binding proteins coexist in eukaryotic cells. *Proc. Natl. Acad. Sci. USA,* **85**, 757.
421. Pfeifer,K., Prezant,T. and Guarente,L. (1987) Yeast HAP1 activator binds to two upstream activation sites of different sequence. *Cell,* **49**, 19.
422. Weintraub,H. (1985) Assembly and propagation of repressed and derepressed chromosomal states. *Cell,* **42**, 705.
423. Grosveld,F., van Assendelft,G.B., Greaves,D.R. and Kollias,G. (1987) Position-independent, high-level expression of the human β-globin gene in transgenic mice. *Cell,* **51**, 975.

3

Termination and 3' end processing of eukaryotic RNA

N.J.Proudfoot and E.Whitelaw

1. Introduction

The transcription of a gene involves three distinct phases: initiation, elongation and termination. While the role of initiation is well known to be of critical importance to the regulation of gene expression (see Chapters 1 and 2), the role of the other two phases of transcription (elongation and termination) is less clearly understood. Elongation may proceed at uneven rates along the gene. Sites at which the RNA polymerase slows down, without release of the RNA transcript, are called pause sites. When termination finally occurs, the RNA transcript is released from the DNA template and in the case of eukaryotes it is rapidly processed. The 3' terminus of the eukaryotic primary transcript rarely corresponds to that of the mature RNA species. In contrast, prokaryotes generate the 3' termini of their RNA by direct transcriptional termination and specific termination signals have been identified at the ends of or sometimes within prokaryotic operons (1). In the simplest cases, a GC-rich hairpin followed by an oligo U sequence in the nascent RNA transcript constitutes a termination signal. Alternatively termination is promoted by a protein factor *rho* at a variety of sequences, all probably associated with transcriptional pausing. The recent discovery that *rho* possesses an ATP dependent RNA:DNA helicase activity suggests that termination is caused by the unwinding of the RNA transcript from the DNA template up to the pause site (2). Another important role of pausing in prokaryotes is in the widespread mechanism of attenuation, in which translation of the primary transcript influences transcriptional pausing and thus the level of mRNA synthesized. This requires the tight coupling of translation and transcription, made possible by the fact that prokaryotes do not segregate their chromosome into a nuclear compartment (see ref. 1 for review).

In eukaryotes only 5S RNA genes appear to generate the 3' termini of their transcripts solely by transcriptional termination (see Section 7.1).

In all other cases investigated, 3′ end formation appears to be a complex combination of both transcriptional termination and RNA processing. The scope of this review is therefore divided into several different parts. We will first review the now substantial weight of experimental data on the 3′ end processing of higher eukaryotic mRNA. We will then consider the case for transcriptional termination in eukaryotic genes transcribed by RNA polymerase II (Pol II genes). While in some genes it is clear that both RNA processing and transcriptional termination are coupled, in other cases it is more likely that termination of transcription is both spatially and mechanistically separate from 3′ end processing. We will then describe what is known about transcriptional termination in RNA polymerase I and III genes. Finally, this surprisingly complex arrangement for the 3′ end formation of eukaryotic RNA has been shown to have a direct and important role in the regulation of eukaryotic gene expression. We review the different types of gene regulation associated with both 3′ end processing and transcriptional termination of eukaryotic RNA in the final two parts of this review.

2. 3′ end processing of polyadenylated messenger RNA

Early experiments on the sequence analysis of a few easily purified mRNAs, such as globin, immunoglobulin and ovalbumin, revealed the first clues to the mechanisms involved in mRNA 3′ end formation. Firstly it was discovered that these eukaryotic mRNAs have a poly(A) tail at their 3′ termini (3–5). Interestingly, when the genes for these mRNAs were isolated by recombinant DNA technology, no poly(dT) tracts were found at the 3′ end of these genes (6). These observations led to the inescapable conclusion that poly(A) is added to mRNA by a post-transcriptional process. Furthermore, poly(A) polymerase activity was detected and the enzyme partially purified from calf thymus nuclei (7). Subsequently the vast majority of eukaryotic mRNA have been shown to possess 3′ terminal poly(A) sequences. Indeed poly(A) has proved an invaluable aid to molecular biologists both as a handle for purification using oligo(dT) cellulose (8) and as a convenient site for priming reverse transcription of mRNA to make cDNA for cloning purposes (9). The exact role of poly(A) in mRNA function is still ill-defined. It is however clear that deadenylated mRNA is relatively unstable (10,11). The efficient addition of poly(A) to mRNA therefore increases the steady-state concentration of mRNA and by such an approach could be used as a mechanism to regulate the expression of a gene (see Section 8.2).

2.1 The AAUAAA sequence
Comparison of the RNA sequences close to the 3′ end of globin, immunoglobulin and ovalbumin mRNAs revealed a striking sequence

homology: AAUAAA. This sequence was found 20–30 bases 5' to the poly(A) tail (12). Subsequently the same sequence or a close homolog has been found in the same 3' terminal position of all polyadenylated mRNAs in higher eukaryotes (13). In fact this sequence represents the most generally universal sequence feature of RNA polymerase II genes and, if nothing else, has proved a useful marker for the 3' end of a gene. The conserved nature of both sequence and position suggested that AAUAAA might have a role in the 3' end formation of mRNA (12), and led to its christening as the 'poly(A) signal' of mRNA. However no direct experiments on the role of AAUAAA were possible until technology for the construction and analysis of mutated AAUAAA sequences was available.

The first analysis of an artificially constructed AAUAAA mutation was carried out on the late gene of SV40 (14). In particular, a 16 bp deletion of sequence including the AAUAAA was found to abolish the 3' end formation of SV40 late mRNA. Furthermore, deletion of sequences between the AAUAAA sequence and the 3' end of the mRNA resulted in an altered placement of the poly(A) tail. The precise mutation of the AAUAAA sequence of adenovirus E1A mRNA to AAGAAA similarly abolished the 3' end formation of this mRNA. AAUAAA was therefore directly implicated in mRNA 3' end formation (15). Interestingly two cases of thalassemia similarly implicate the AAUAAA sequence in the 3' end formation of mRNA. In one case of α^+ thalassemia, the AAUAAA is mutated to AAUAAG (16) while in one case of β^+ thalassemia AAUAAA ia altered to AACAAA (17). Both of these point mutations result in elongated globin mRNAs extending to cryptic poly(A) sites some distance into the 3' flanking sequence of the gene. In the case of the α^+ thalassemia mutation, the cryptic poly(A) site used instead of the normal site is apparently an inefficient signal so that reduced levels of mRNA are obtained, in part explaining the phenotype of the mutation (18). Many other variants of the ubiquitous AAUAAA sequence have now been synthesized and tested (19). Most abolish 3' end formation while others reduced the efficiency of the process. Indeed the possession of variant AAUAAA sequences in naturally occurring genes may reflect the intentional possession of a weak poly(A) site as a simple approach to reducing the amount of mRNA produced by a particular gene (see Section 8.2).

2.2 Downstream poly(A) site signals

The possibility that additional sequences to AAUAAA may play a role in mRNA 3' end formation arose from the fact that perfect copies of the AAUAAA sequence can occasionally be found within genes at positions where no mRNA 3' ends are apparently formed. Similarly many genes possess multiple copies of AAUAAA sequences towards the 3' end of the mRNA which may each elicit 3' end formation to variable extents

(20 – 22). A number of studies have therefore analyzed sequences close to the AAUAAA sequence of different mRNAs which might play an additional role in 3' end formation (23 – 30). In several different cases such sequences have been defined immediately 3' to the end of the mRNA in the 3' flanking region of the gene. Thus the deletion of these downstream sequences abolishes 3' end formation. Careful dissection reveals that either a GT-rich or T-rich genomic sequence element may constitute the additional signal. Indeed extensive computer analysis of the 3' ends of Pol II genes reveals the general although not universal possession of either or both a GT- and T-rich sequence (30). Recent studies have further demonstrated that the position of the corresponding GU-rich and U-rich downstream sequences in mRNA with respect to the AAUAAA sequence is precise, so that abnormal separations of the two elements abolishes the 3' end formation of mRNA (31 – 33). The absolute requirement for GU-rich and/or U-rich sequences downstream of some poly(A) sites raises the paradox of how other natural poly(A) sites apparently get by without these sequence elements. Indeed, in the case of *Xenopus* β-globin gene expression in *Xenopus* oocytes, apparently no specific downstream sequence elements are required (34). These differences in sequence requirements may reflect differences in the detailed mechanism of mRNA 3' end formation in different cell types, although no direct evidence for this is available.

2.3 *In vitro* 3' end processing and polyadenylation

The identification of sequence elements at the 3' end of a gene which are associated with mRNA 3' end formation in no way distinguishes the two possible mechanisms by which such 3' end formation could occur: RNA processing or transcriptional termination. However, convincing evidence for the 3' end processing of mRNA comes from *in vitro* polyadenylation experiments. The first *in vitro* polyadenylation activity specifically associated with mRNA 3' ends involved the addition of poly(A) tails to RNA templates synthesized *in vitro* that had 3' termini just beyond an AAUAAA sequence (35). This activity was found in whole cell extracts of HeLa cells. Subsequently both an endonuclease and polyadenylation activity was demonstrated in nuclear extracts (36,37). RNA templates [synthesized *in vitro* with bacteriophage SP6 RNA polymerase (38)] extending past the 3' end of an adenovirus poly(A) site were correctly cleaved and the newly formed 3' ends polyadenylated. This experiment formally proved that mRNA 3' ends are generated by an RNA processing mechanism. The fact that the nuclease activity was an endonuclease rather than an exonuclease was demonstrated by the identification of the 3'-side RNA fragment of the processed RNA template. Furthermore the endonuclease and polyadenylation reaction could be uncoupled by inhibiting polyadenylation using analogs of rATP, such as cordycepin (3' deoxyATP). In the presence of cordycepin, cleavage, but not polyadenylation, occurs at full efficiency. The sequence specificity of these *in vitro* 3' end processing activities closely matches the *in vivo* sequence requirements. Thus both the AAUAAA sequence (39) as well as the downstream GU- and U-rich sequence elements are required (40,41).

The detailed biochemical analysis of the factors involved in 3' end processing is currently under way. Following the precedent of RNA splicing, it is widely thought likely that at least one component in the polyadenylation complex will be a small nuclear RNA together with its associated RNA binding proteins (snRNPs) (42). Indeed snRNPs have been directly associated with polyadenylation in a number of studies (36,43). Thus *in vitro* 3' end processing and polyadenylation can be inhibited with antibodies directed against snRNPs (36) and the RNA sequences precipitated by snRNP antibodies can be demonstrated to include the poly(A) sites of mRNA (43). However, the exact nature of the snRNP involved is, at the date of writing this review, unknown. One serious problem in snRNP identification is that, since polyadenylation of eukaryotic mRNA is a major process in mRNA maturation, it might be expected that one of the predominant small nuclear RNAs would have a direct role. However, of the six high yield snRNPs (U1 – U6) the U1, U2, U4 and U6 RNAs are known to be directly involved in intron splicing (45) and are not required for 3' end processing (43,44). Similarly, U5 is required for intron splicing (45) while U3 is implicated in ribosomal RNA maturation (46). It is difficult to see how a predominant snRNP could have an exclusive role in polyadenylation.

Our current state of knowledge of 3' end formation for polyadenylated mRNA is summarized in *Figure 1*.

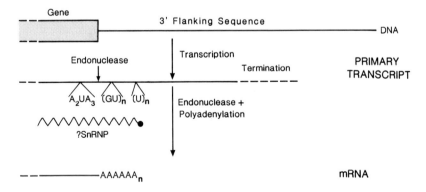

Figure 1.   Formation of the 3' end of polyadenylated mRNA.

3. 3′ end processing of histone messenger RNA

Histone genes in eukaryotes have a number of unique features that distinguish them from other RNA polymerase II genes. In most species investigated, they are repetitive genes often clustered into repeat units with each unit containing one copy of the five genes, H1, H2A, H2B, H3 and H4 (47). Furthermore, these genes do not contain introns and their mRNAs are not polyadenylated (except in exceptional cases) (48).

3.1 Signals for histone mRNA 3′ end formation

The first clues as to the mechanism of histone mRNA 3′ end formation again came from a comparison of the 3′ terminal sequences of different histone genes. Nearly all histones have a conserved 3′ end to their mRNA sequence, ACCA (49). Superimposed over this absolute sequence motif is a wider consensus sequence of about 20 nucleotides which is invariably an inverted repeat structure. A second region of sequence conservation, CA_2GA_3GA, is also invariably present in the immediate 3′ flanking region of the histone gene (49). Both regions of sequence conservation are directly implicated in histone mRNA 3′ end formation. Deletion and point mutation analysis of the sea urchin H2A gene sequence followed by injection of the mutant H2A genes into *Xenopus* oocytes showed that both the 3′ terminal inverted repeat sequence and the immediate downstream element were essential for 3′ end formation (50–52). Furthermore analysis of point mutations in the inverted repeat sequences as well as revertant mutations that maintained mRNA base pairing but altered the actual nucleotide sequence revealed that the capability of the histone gene transcript to form a hairpin loop at the 3′ end of the mRNA was essential for mRNA 3′ end formation (51).

3.2 The role of U7 snRNP

It was discovered that the sea urchin H3 gene was incapable of mRNA 3′ end formation when microinjected into *Xenopus* oocytes. However it did form authentic mRNA 3′ termini when nuclear extracts isolated from sea urchin nuclei were co-injected with the H3 gene. This approach in effect provided a classic complementation assay (53). Following an extensive fractionation of sea urchin extracts, relatively pure fractions were found to elicit 3′ end formation of H3 mRNA when co-injected with the H3 gene. The active component of this positive fraction was found to be a hitherto uncharacterized small nuclear RNA, U7 (54). The association of snRNPs with the RNA processing of mRNA (45) immediately suggested a 3′ end processing mechanism for histone mRNA maturation as well. Indeed soon after, these experiments were carried out on RNA templates synthesized *in vitro* which extended past the 3′ end of various histone mRNAs into their 3′ flanking regions. Incubation of these RNA 'substrates' with nuclear extracts (55,56) or injection into

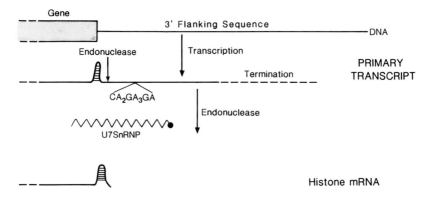

Figure 2. Formation of the 3' end of histone mRNA.

Xenopus oocytes (57) resulted in 3' end processing to generate histone mRNA with authentic 3' termini. Furthermore the same 3' terminal hairpin and 3' flanking CA_2GA_3GA sequence were absolutely required for this RNA processing activity (52,56). Again, as with polyadenylated mRNA, this 3' processing activity was shown to be an endonuclease (58).

U7 RNA has subsequently been sequenced by cloning its cDNA transcript. Interestingly, significant regions of homology between the U7 sequence and the histone mRNA 3' processing signals are apparent (59). This suggests that U7-mediated 3' processing involves RNA:RNA base pairing between U7 and the histone precursor RNA. The cloning of the gene encoding U7 RNA has allowed an unequivocal test for RNA:RNA base pairing. Point mutations in the 3' flanking CA_2GA_3GA sequence of the histone H3 gene were matched by the construction of compensatory point mutations in the U7 RNA gene that reformed base pairs albeit with a different sequence. The mutant H3 gene that was demonstrated to lack the ability to 3' process when injected into *Xenopus* oocytes could be demonstrated to 3' process efficiently when co-injected with the compensatory U7 RNA mutant gene. These definitive experiments prove that U7 RNA directly base pairs with the CA_2GA_3GA sequence in the H3 mRNA precursor (60,61).

The overall scheme of histone mRNA 3' end formation is shown in *Figure 2*.

4. 3' end formation of small nuclear RNAs

The genes encoding the snRNAs U1–U7 are transcribed by RNA polymerase II (62), except for U6 which may be a Pol III gene (63). As with histone genes, snRNA genes are short in length, lack introns and encode RNA species which lack poly(A) tails. Again, as with histone genes, snRNA genes have a hairpin loop sequence at the 3' end of their mature RNA although this sequence is not especially conserved between snRNAs. Similarly a conserved sequence, $GU_3A_3N_3AGA$, is present in the immediate 3′ flanking sequence of U1 and U2 snRNA genes and is reminiscent of the histone CA_2GA_3GA 3′ processing signal. Deletion analysis of snRNA genes has revealed that the conserved 3′ flanking region sequence is absolutely required for the 3′ end formation of snRNA (64,65). In contrast the 3′ terminal hairpin loop sequence is not required for 3′ end formation but may influence the stability of the snRNA (66).

Nuclear run-off analysis on the U1 gene (see Section 6.1) suggests that transcription terminates soon after the conserved 3′ flanking region sequence (67). Recently it has also been demonstrated that snRNA 3′ end formation is coupled to the initiation of transcription (68,69). Replacement of the snRNA promoter with a heterologous Pol II promoter abolishes authentic snRNA 3′ end formation. This result argues that factors required for 3′ end formation are present on the initiation complex of the snRNA gene. Finally a number of unpublished studies (cited in refs 64 and 65) have unsuccessfully looked for a 3′ end processing activity by making *in vitro* RNA transcripts of snRNA 3′ ends that extend into the 3′ flanking sequence. Incubation of these RNAs with nuclear extracts under conditions known to allow 3′ end processing of polyadenylated mRNA gave no detectable 3′ end processing. Taken together, the inability to detect a 3′ processing activity with the apparently close association of transcriptional termination to the 3′ end of U1 snRNA suggests that small nuclear RNA genes generate their 3′ termini by direct transcriptional termination. A small amount of 3′ exonuclease activity may tidy up the primary terminated RNAs to the base of the snRNAs 3′ terminal hairpin loop structure. Clearly the conserved 3′ flanking region homology may be a transcriptional termination signal although other signals may also be involved in the termination process.

Figure 3 summarizes our current understanding of 3′ end formation for snRNA.

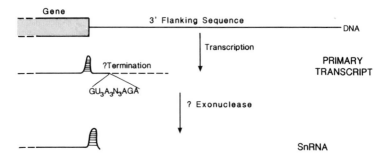

Figure 3.   Formation of the 3′ end of SnRNA.

5. 3′ end formation of yeast mRNA

Yeast mRNA is polyadenylated as in higher eukaryotes but lacks the expected AAUAAA sequence motif. This fact suggests that yeast may have a different mechanism for the 3′ end formation of mRNA. A small deletion mutant in one of the cytochrome *c* genes of yeast (CYC1) provided direct proof of yeast's individuality (70). This mutant gene lacks 38 bp close to the 3′ end of the gene. As a result, low levels of mRNA were detected with 3′ termini extending to different positions into the 3′ flanking region. All of these aberrant mRNAs were polyadenylated. Another important study (71) revealed that when the *Drosophila* ADE-8 gene or deletions thereof were transfected into yeast, poly(A) addition occurred at sites close to the consensus sequence TTTTTATA. Furthermore this same consensus sequence is found in the CYC1 deletion. However a number of different yeast mRNAs do not possess obvious versions of this consensus sequence. Since nuclear run-off analysis (see Section 6.1) has not been reported for yeast Pol II genes, it has proved difficult to distinguish 3′ end processing from termination of transcription as the mechanism for the 3′ end formation of yeast mRNA. However, some studies (72) on yeast mRNA have favored the notion that transcriptional termination is closely coupled to polyadenylation with no obvious role for mRNA 3′ end processing in mRNA maturation.

6. Termination of transcription in genes transcribed by RNA polymerase II

The discovery that the 3′ end formation of eukaryotic mRNA is carried out by an RNA processing mechanism was unexpected. Furthermore the unanswered question posed by these results is where does transcription terminate if not at the 3′ end of the mRNA? One possibility could be that, since mRNA 3′ ends are formed by RNA processing, perhaps termination is a wholly uncontrolled, random process. Termination occurring before the 3′ processing signals would result in unstable transcripts turned over in the nucleus while termination occurring after the 3′ processing signals would be tidied up into authentic mRNA 3′ ends by RNA processing. The simplicity of this random approach to termination is clearly attractive as no additional genetic complexity would be required. However a number of considerations make this hypothesis untenable. Since RNA polymerase II genes vary in length enormously from as little as 0.5 kb [as with a histone gene (44)] to 100 kb [as with the clotting factor VIII and IX genes (73,74)], the random termination model would preclude the efficient transcription of larger genes. Furthermore the possibility of transcriptional interference between genes (see Section 9.2) argues for the existence of an efficient termination process. These considerations have led a number of laboratories to search for authentic transcriptional termination signals in or beyond Pol II genes. However, these studies are inevitably hampered by the extreme instability of primary RNA transcripts.

6.1 Analysis of termination by nuclear run-off

The technique of nuclear run-off analysis has been developed specifically to cope with the difficulty in analyzing unstable primary transcripts from a gene (75). The level of nascent RNA transcription can be measured by the labeling of the transcript as it is synthesized, uncoupled from the relative stability of different RNA transcripts. Unfortunately the method is difficult to carry out unless the gene to be analyzed is transcribed at very high rates. Furthermore the technique is effectively an *in vitro* transcription procedure, so that the results obtained may not be wholly physiological. The procedure is as follows.

Nuclei are prepared from cells in which the gene being analyzed is active. These nuclei are incubated in a transcription buffer containing $[\alpha\text{-}^{32}P]UTP$ for a short period. During this incubation period, pre-initiated RNA polymerase molecules will incorporate $[\alpha\text{-}^{32}P]UTP$ into the elongating primary RNA transcripts. In some studies the length of 'nuclear run-off' transcript added to the primary RNA has been estimated as only a few hundred nucleotides, presumably due to the low concentration of UTP in the transcription buffer. Apparently the rate of transcription is very low in these nuclear run-off experiments. Radioactive nuclear RNA labeled by this procedure is purified from the nuclei and hybridized to DNA fragments corresponding to different parts of the gene to be analyzed.

Quite a large number of different eukaryotic RNA polymerase II genes have now been analyzed by the nuclear run-off assay (see *Table 1*). In all cases, equimolar transcription is reported through the gene and extends

Table 1.   Termination sites for eukaryotic RNA polymerase II genes

Gene	Distance beyond poly(A) site (bp)	Reference
α_D-globin (chicken)	~600	81
α_A-globin (chicken)	~1500	81
α-globin (mouse)	100 – 300	82
α-globin (human)	100 – 300	18
β-globin (mouse)	600 – 1800	76,77
β-globin (rabbit)	1000 – 1700	80
DHFR (mouse)	1000 – 2000	83,84
α-amylase (mouse)	2000 – 4000	78
ovalbumin (chicken)	800 – 1000	79
calcitonin/CGRP (rat)	1000 – 2000	85

for a surprisingly variable distance into the 3' flanking region ranging from a few hundred nucleotides to many kilobases. For some genes the cessation of transcription in the 3' flanking sequence would appear to be a rather imprecise process with nuclear run-off signals diminishing gradually over a significant distance (~ 1 kb), such as with mouse β-globin (76,77) and α-amylase (78). Other genes appear to have a more precise termination site within only $100-200$ bp, such as chicken ovalbumin (79). A major drawback to analyzing transcriptional termination by nuclear run-off analysis is that the precise 3' termini of the transcript cannot be identified because transcription is only indirectly detected by hybridization to genomic DNA fragments immobilized on filters.

6.2 Termination signals

Recently a number of studies have begun the process of defining transcriptional termination signals. In each case putative termination signals have been manipulated in bacterial plasmids and re-inserted into the middle of a usually heterologous transcription unit. In the case of the mouse β-globin gene, the 3' flanking region including the region implicated in transcription termination was inserted into the adenovirus E1A gene using a recombination procedure (86). By this approach a number of different constructs were tested which revealed that the termination process was orientation-specific and apparently required the 3' terminus of the mouse β-globin gene as well as the 3' flanking sequence. The more precise definition of required signals has not so far been reported. Similarly the termination process for the human gastrin gene has been extensively characterized and has been shown to involve a T-rich sequence in the immediate 3' flanking region of the gene (87,88). It should be noted, however, that this T-rich sequence might have an alternative role in mRNA stability (see Section 8.1).

The sequences associated with transcriptional termination of the sea urchin H2A histone gene have also been studied (89). It was discovered that although the sea urchin histone gene does not 3' end process when transfected and expressed in human HeLa cells, it still prevents transcriptional readthrough when inserted into the human α-globin gene which probably relates to an active termination process. The sequences required for this 'termination process' were dissected and found to comprise three separate sequence elements. One was within the coding sequence of the histone gene while the second was an A-rich sequence a few hundred nucleotides past the 3' processing signals. Finally a third region was required towards the end of the 3' flanking sequence close to the 5' end of the adjacent H1 histone gene. Comparing the gastrin and H2A histone 'terminators', it is striking that a long stretch of A or T residues are in both cases implicated in the process. Much more analysis of both the signals and factors involved in these processes will be required before a more complete understanding of transcriptional termination is available. Indeed it will prove necessary to develop *in vitro* transcription systems that authentically terminate transcription so that a full biochemical analysis of termination can be achieved.

6.3 Coupled transcriptional termination and mRNA 3′ end formation in RNA polymerase II genes

A plausible mechanism for Pol II termination of transcription could be that 3′ processing of the primary transcript occurs soon after the passage of RNA polymerase past the 3′ end of the mRNA. This processing event would expose a 5′ end on the primary transcript close to the elongating polymerase. Possibly through the action of exonucleases and/or an RNA: DNA helicase (2), the transcript immediately adjacent to the polymerase could then be rapidly released and degraded, thereby dissociating the polymerase and effectively promoting termination of transcription. Such a mechanism would be augmented by either the pausing of the polymerase on the gene, which would allow time for the processing events to elicit termination, or by the possession of a very efficient poly(A) site which would promote 3′ end processing immediately the RNA polymerase passed the poly(A) site.

Such a mechanism has been suggested by studies on the human α2-globin gene (18). We have demonstrated by nuclear run-off analysis that transcription termination occurs soon after RNA polymerases pass the poly(A) site of the α-globin gene. However a point mutation in the AAUAAA sequence (derived from a type of α^+ thalassemia) not only fails to 3′ end process but also fails to terminate transcription. The close proximity of transcriptional termination to the poly(A) site of the α-globin gene may well be related to this apparent coupling of the two processes. However, the fact that α-globin gene termination appears to require a functional poly(A) site does not exclude the possibility that there are additional specific signals associated with transcriptional termination or pausing. A second case relevant to this mechanism may be that of the mouse β-globin gene. Although termination occurs over 1 kb past the 3′ end of the mRNA (77), the poly(A) site region is still required for efficient termination of transcription (86).

7. Transcriptional termination in RNA polymerase III and I genes

7.1 RNA polymerase III transcriptional termination

Transcriptional termination in the RNA polymerase III genes, 5S and tRNA appears to be a relatively simple process. The 3′ terminus of *Xenopus* 5S RNA is generated by a direct transcriptional termination process, apparently requiring no additional factors to the elongating RNA polymerase III complex. *In vitro* transcription of the 5S gene with purified

RNA polymerase III generates 5S RNA with authentic 3' ends (90). The termination signals controlling this process are invariably short runs of T residues, at least four, and are usually found surrounded by GC-rich sequences (91). The 3' termini of Pol III transcripts are therefore invariably oligo U sequences. The tRNA genes again require short runs of T residues for transcriptional termination. However the sequence environment of the oligo T sequence appears to be important in that, with some tRNA genes, oligo T sequences within the gene do not elicit transcriptional termination (92).

The primary transcripts of tRNA genes are subsequently extensively processed to yield the mature tRNA molecule. The formation of tRNA 3' ends involves both the 3' end processing of the primary transcript and the post-transcriptional addition of a CCA trinucleotide (see ref. 93 for review of tRNA processing).

7.2 RNA polymerase I transcriptional termination

In contrast to RNA polymerase III transcriptional termination, RNA polymerase I termination is a very much more complex affair and is only now becoming more fully understood. RNA polymerase I (Pol I) genes appear to have evolved every possible trick for high transcriptional output. Pol I genes are localized to specific sites within the nucleus called nucleoli. They are transcribed by their own private RNA polymerase. They are a highly repetitive gene family of tandemly repeated gene units (94). Finally, at least in some species, they possess multiple promoter-like elements which appear to enhance the level of transcriptional initiation (95).

Transcriptional termination in Pol I genes was for many years assumed to occur at the 3' terminus of the 28S rRNA gene. Indeed the elegant electron micrographs of so-called 'Christmas tree structures' over the gene regions of the ribosomal DNA repeat unit lent considerable weight to these arguments (96). No nascent RNA tails are visible past the 3' terminal region of the 28S rRNA gene on these electron micrographs. However, as described below, the physiological significance of this picture is now open to question. More recently, direct evidence for transcription past the 3' end of the 28S rRNA gene and into the 3' flanking spacer sequence has been described in the mouse. Both nuclear run-off analysis and *in vitro* transcription experiments have pinpointed a termination site 500 bp past the 3' end of the 28S rRNA gene (97). This termination site corresponds to a short sequence containing a *Sal*I restriction enzyme site repeated eight times in all (T1 – T8 in *Figure 4A*). It is interesting that a specific protein factor has now been demonstrated to bind to each of these repeated 'Sal boxes' and is thus inferred to be a transcriptional termination factor (98). The 3' terminus of the 28S rRNA must therefore be formed by an RNA processing mechanism (*Figure 4A*). Indeed this processing event must occur very early in the maturation of rRNAs since

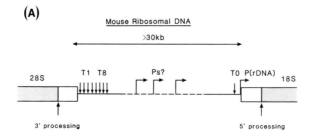

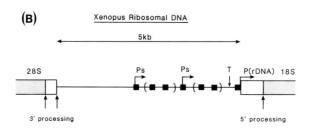

Figure 4.   The spacer sequence of (**A**) the mouse and (**B**) *Xenopus* ribosomal gene repeat unit. T denotes terminator sequence. Ps denotes spacer promoter and P(rDNA) denotes the major Pol I promoter with the arrows indicating the direction of transcription. Hatched boxes denote 28S and 18S RNA coding sequence. Open boxes denote precursor rRNA sequence. Horizontal lines denote spacer sequence.

the earliest accumulating precursor rRNA is the 45S RNA product which has a 3′ end co-terminal with 28S rRNA. The mechanism of 28S rRNA 3′ end formation is largely unknown. Interestingly, U3 snRNA has been implicated in rRNA maturation, raising the possibility that rRNA 3′ end formation could be yet another example of an snRNP-mediated process (46).

7.3 Transcriptional enhancement through RNA polymerase I termination

A number of intriguing observations have recently demonstrated that Pol I transcriptional termination is a still more complex phenomenon.

7.3.1 Drosophila and Xenopus rDNA

In both *Drosophila* (99) and *Xenopus* rDNA (100) repeat units, transcription can be detected in the spacer sequence between two adjacent gene units. Indeed, based on nuclear run-off analysis of *Xenopus* rDNA, it is clear that transcription initiating on the major promoter 5′ to the 18S rRNA sequence reads through not only the 45S precursor region but all of the way through the spacer sequence to within 200 bp of the promoter for the next repeat unit (101). At this point there is a clearly defined transcriptional terminator sequence (*Figure 4B*). Mutations in this region eliminate transcriptional termination. The discovery of this terminator sequence right up against the next promoter neatly explains what happens to transcription off the multiple promoters shown to exist in the rDNA spacer sequences of some such species such as *Xenopus* (95). These spacer promoters do indeed fire transcription towards the next ribosomal gene repeat unit (*Figure 4B*) but as with the transcripts that readthrough from the downstream ribosomal gene, this additional transcription also terminates at the promoter-proximal terminator sequence. Indeed the close proximity of this terminator to the major Pol I promoter appears to be essential for efficient Pol I transcription (102). Either deletion of the terminator sequence or the artificial separation of it from the promoter significantly down regulates transcription. These data illustrate an ingenious approach to efficient transcription. RNA polymerase I is effectively concentrated onto the major promoter because it is only released from the RNA template directly adjacent to where it must re-initiate on the next promoter. The existence of multiple promoters in the spacer sequence provides additional RNA polymerase I traps for any enzyme that is not already transcribing a previous rDNA repeat unit (95).

7.3.2 Mouse rDNA

The spacer sequence of *Xenopus* and *Drosophila* rRNA genes is relatively small, only about 5 kb. It would therefore seem that the apparent transcriptional wastage of transcribing whole spacer sequences is outweighed by the benefit of increasing the efficiency of initiation of transcription. However, with mammalian ribosomal genes the spacer is much larger, about 30 kb in mouse. It therefore seems sensible that efficient termination occurs fairly soon after the 3' end of the 28S RNA gene as described in Section 7.2. Even so it is interesting to note that in mouse there is an additional Sal box terminator sequence (T0) next to the major promoter (*Figure 4A*). As with *Xenopus*, this terminator encourages more efficient transcription off the major promoter (103,104). Presumably, as with *Xenopus*, there are multiple promoters in the mouse rDNA spacer which funnel RNA polymerase via the promoter-proximal terminator sequence onto the next promoter. However, these presumptive spacer promoters have not yet been identified in mammalian ribosomal genes.

8. Regulation of gene expression by mRNA 3' end processing

This review documents a surprising level of complexity for the, at first sight, simple business of forming the 3' termini of eukaryotic RNA molecules. It seems a reasonable assumption that the considerable number of extra genes required for these complex processes, such as the genes encoding the endo- and exonucleases, the poly(A) polymerase, the snRNAs and associated factors and possible termination factors, are there for an important biological purpose. It seems plausible to us that such a purpose may relate to the regulation of gene expression.

8.1 mRNA stability

A simple approach to regulate the output from a gene is by controlling the stability of the gene transcript. This approach is not unique to eukaryotes since it has been demonstrated that prokaryotes can significantly modulate their mRNA levels by increasing mRNA stability. The case of the bacterial REP sequences, a repetitive inverted repeat element, is a clear illustration of this point. These sequences are found at the 3' end of a large number of bacterial mRNAs and appear to stabilize mRNA sequences on their 5' side by blocking the action of the predominant 3'→5' exonuclease activity (105). The 3' terminal hairpin sequences of histone mRNA and snRNAs are highly reminiscent of the prokaryotic REP sequences and very possibly protect their 3' termini from exonuclease degradation by a similar mechanism.

Another example of a mechanism for modulating mRNA levels as a means of controlling gene output comes from the class of eukaryotic mRNAs that are associated with only very transient expression such as interferon, c-*myc*, c-*fos* and various growth factor mRNAs. All of these mRNAs have AU-rich sequences in their 3' non-coding regions with several copies of the sequence AUUU in tandem. A recent study (106) clearly demonstrates that this AU sequence confers instability on the mRNA. The AU-rich sequence from the 3' non-coding region of the granulocyte macrophage colony stimulating factor (GM-CSF) gene was synthesized *in vitro* and cloned into the 3' non-coding region of the rabbit β-globin gene. As predicted, the insertion of this sequence rendered the otherwise highly stable rabbit β-globin mRNA extremely unstable. In fact it is worth noting that the putative gastrin gene terminator (87) is a very similar sequence containing multiple AUUU sequences. It remains a possibility that this sequence is involved in RNA instability rather than termination of transcription.

The possession of a poly(A) tail is another clear example of a sequence that confers stability on an otherwise unstable mRNA molecule (for review, see ref. 107). In some now classic experiments, the poly(A) tail of an α-globin mRNA was precisely removed by oligo(dT)-directed RNase H digestion. The deadenylated mRNA was injected into *Xenopus* oocytes (10). While the normal α-globin mRNA was stable and translationally competent for up to a week, the deadenylated mRNA disappeared after only a few hours. More recently, using the SP6 *in vitro* transcription system (38), it has proved possible to synthesize mRNA with or without poly(A) tails. The injection of these mRNAs into *Xenopus* oocytes revealed that polyadenylation results in increased stability (11). It therefore follows that the efficiency of polyadenylation could regulate the level of mRNA. Indeed it has recently been shown that normally unstable, antisense transcripts originating within the murine c-*myc* gene are stabilized following certain chromosomal translocations which provide them with a functional poly(A) site (108). This provides an *in vivo* example of the role of polyadenylation in mRNA stability.

8.2 Efficiency of an mRNA poly(A) site

It is intriguing to speculate that the design of each gene's poly(A) site is intentionally adjusted to provide the appropriate level of mRNA for that gene. A case in point may be AUUAAA, the relatively common variant of the AAUAAA sequence [10% of poly(A) sites] as well as the two less common variants, AAUACA and AAUUAA [~2% of poly(A) sites]. In the latter two cases it has been directly demonstrated that these poly(A) sites are considerably less efficient than the normal AAUAAA sequence (19). Clearly, genes possessing such variant poly(A) sites may end up producing lower steady-state mRNA levels. Of course, such a prediction would depend on whether or not poly(A) addition is a rate-limiting step in mRNA synthesis.

The additional GT- or T-rich downstream sequences required for polyadenylation may have a similar role in the modulation of mRNA levels. Indeed the fact that some genes have convincing GT- or T-rich sequence elements while others do not may directly relate to the efficiency of their poly(A) sites. This hypothesis has been tested by reconstructing the poly(A) site of the rabbit β-globin gene (32). To do this, synthetic oligonucleotides containing either the GT- or T-rich sequences were added to a deleted non-functional form of the rabbit β-globin gene from which these downstream sequences had been deleted. It was found that both the GT- and T-rich elements were required in their wild-type spatial relationship with respect to the end of the mRNA. Either the absence of one downstream element or their altered position greatly reduced the levels of mRNA produced by the gene. Thus, it is possible that genes with variant or non-existent downstream poly(A) site signals will produce significantly reduced levels of mRNA.

8.3 Differential utilization of poly(A) sites

It is now clearly documented that the mechanism of alternative splicing or the joining of different combinations of exons is an important way of increasing the diversity of proteins made by one gene (109). Several genes such as tropomyosin or fibronectin make extensive use of this process. This section documents the as yet rare but probably very important process of alternative poly(A) site selection in some genes as another approach to increasing the diversity of proteins made from one gene.

Many eukaryotic genes have been found to synthesize multiple mRNAs differing in the lengths of their 3' non-coding sequences (20–22,110–112). Sequence analysis of the 3' non-coding region of these genes reveals multiple poly(A) sites. The rules governing which of the multiple poly(A) sites is predominantly utilized are unknown. Presumably the most efficient poly(A) site predominates. The fact that mRNAs do not invariably utilize the most 5' positioned poly(A) site demonstrates that once one poly(A) site has been selected and polyadenylation takes place, no further 3' end processing of 5' positioned poly(A) sites can occur.

When multiple poly(A) sites are present which are not exclusively within a gene's 3' non-coding region but in different exons of a gene, alternate forms of protein can be made. Two examples of this phenomenon are well documented: the immunoglobulin constant region gene (113–115) and the calcitonin gene (116). In both cases, the different protein forms are specific to different cell types, arguing that factors within these different cells regulate the selection of a particular poly(A) site. *Figures 5* and *6* outline the arrangement of these two gene systems. For the immunoglobulin μ constant region (and probably other classes of constant region), the utilization of the two different poly(A) sites indicated in *Figure 5* results in either the C-terminal secreted form of the protein or the larger

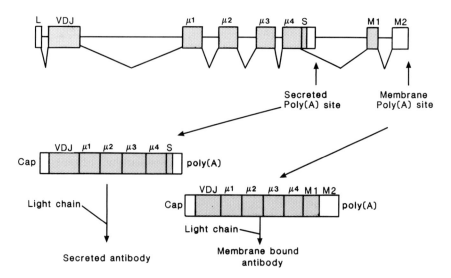

Figure 5.   Differential poly(A) site selection of the immunoglobulin constant region. The lettering of the exons is as follows: L denotes leader exon, V denotes variable, D denotes diversity and J denotes joining portions of the VDJ exon. μ1–μ4 denote the four μ constant region exons. S denotes the secreted form C terminus whereas M1 and M2 are the two membrane form exons. The positions of the secreted and membrane poly(A) sites are indicated by vertical arrows. Open boxes denote non-coding exons, while hatched boxes denote coding exons. Horizontal lines denote introns. Lines joining exons below the gene map indicate splicing reactions.

membrane-bound form of the protein. In certain B cells only the secreted form is made, while in other B cells the membrane-bound form is made. It has been a matter of controversy as to what mechanism directs the selection of these two poly(A) sites. It is formally possible that termination of transcription could occur between the secreted (S) and membrane exons (M1, M2) in B cells that express only secreted antibody (*Figure 5*). However nuclear run-off analysis shows that transcription proceeds past the 3' end of the membrane exons even in cells synthesizing secreted antibody (117). Furthermore, in some B cell lines expressing only the secreted form, RNA fragments have been detected corresponding to the membrane exons (118). These results formally exclude termination as the mechanism defining poly(A) site selection.

It is possible that splicing the μ4 exon to the M1 exon is cell-specific and occurs before poly(A) addition. In other words, if splicing occurs first, then the membrane poly(A) site must be used; if not, the secreted poly(A) site can be utilized. Alternatively polyadenylation may precede splicing, with the differential recognition of the two poly(A) sites determining which form of protein is synthesized. Strong evidence for this latter mechanism exists in that deletion of the secreted poly(A) site results in the production

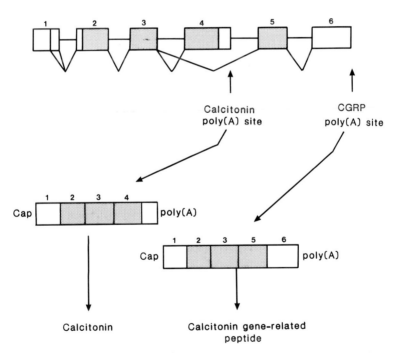

Figure 6. Differential poly(A) site selection of the calcitonin gene. 1–6 are separate exons. Open boxes denote 5' and 3' non-coding sequence exons; hatched boxes denote coding sequence exons. Horizontal lines denote introns. Lines joining exons below the gene map indicate splicing reactions. The splicing of exon 1 to 2 occurs at two alternative donor sites.

of mRNA utilizing the membrane poly(A) sites in B cell lines which normally only synthesize secreted antibody (119). In other words it appears to be the possession of the secreted poly(A) site that controls the synthesis of secreted antibody.

Interestingly, the opposite mechanism appears to operate for the calcitonin gene (116). As shown in *Figure 6*, this gene has two poly(A) sites. The first one [the calcitonin poly(A) site] is at the end of exon 4 and is predominantly utilized in thyroid tissue. Alternatively, in various brain tissues, exon 4 is spliced out so that exons 1, 2 and 3 are joined to two further exons 5 and 6 ending in the CGRP (calcitonin gene-related peptide) poly(A) site. Again nuclear run-off analysis suggests that termination of transcription occurs 3' to both poly(A) sites so that termination cannot control which poly(A) site is utilized (85). However in this gene system inactivation of the 5' calcitonin poly(A) site does not cause a commensurate utilization of the 3' CGRP poly(A) site in cells that normally only synthesize calcitonin. Rather, unspliced precursors accumulate in the nucleus seemingly unable to process to either mRNA form (120). These data therefore argue that thyroid tissue lacks the ability to splice out exon 4 to join exon 3 to exon 5 so that in the absence of a functional calcitonin poly(A) site no stable mRNA can be produced by this gene. The obvious versatility of differential poly(A) site selection as a mechanism for increasing the diversity of proteins produced by a gene as described in these two cases raises the possibility that this mechanism may occur in other as yet uncharacterized gene systems.

8.4 Histone mRNA 3' end processing and cell cycle regulation

It has been clear for some time now that histone mRNA concentrations are carefully regulated during the cell cycle (48). Apparently it is essential to the viability of a cell that the concentrations of histone mRNA and hence histone proteins are tightly coupled to DNA replication. This regulation is achieved at several different levels. Firstly it has been demonstrated that histone promoters fire transcription at very different rates during the cell cycle, with the highest rate of transcription initiation occurring at the end of the G1 phase and the beginning of the S (replication) phase (121,122). It has also been shown that histone mRNA stability is an important factor in the cell cycle regulation of histone mRNA concentration. Histone mRNA is most stable during the S phase of the cell cycle and is destabilized following the cessation of replication (121).

Recent experiments have revealed the intriguing fact that the signals associated with histone mRNA 3' end processing are also associated with cell cycle regulated stabilization (123,124). Indeed it seems likely that 3' processing of histone mRNA may occur in a synchronized manner to generate mRNA only during the S phase. Evidence for this prediction came from experiments in which the 3' processing signals of a histone gene were inserted into the SV40 early gene. mRNA initiating on the SV40 early promoter and ending on the histone 3' processing signals was clearly cell cycle regulated in that it was present at highest levels during the S phase. Interestingly mRNA reading through the histone 3' processing signals to the SV40 early gene poly(A) site was present at equal levels throughout the whole cell cycle (123). This argues that it is the 3' processing event of the histone mRNA rather than specific signals within the histone mRNA that confers cell cycle regulation. Alternatively these specific signals must be at the 3' end of the mRNA to exert their effect on RNA stability.

   The final level of control exerted on histone mRNA appears to be the controlled destabilization of the mRNA at the end of the S phase. It has been known for some time that histone mRNA becomes destabilized following the cessation of replication and that this destabilization can be prevented by inhibiting protein synthesis (125 – 127). Again, recent studies have revealed that the 3'-terminal hairpin loops of the histone mRNA (Section 3.1 and *Figure 2*), as well as its distance from the termination codon of the mRNA, is critical to the destabilization process (128). Artificial histone genes have been constructed in which the 3'-terminal hairpin loop of the mRNA is moved further away from the translated portion of the mRNA either by expanding the 3' non-coding region of the mRNA or by introducing stop codons within the coding sequence. Both histone gene constructs produce mRNA with significantly increased stability. These results re-affirm the connection between translation and destabilization previously suggested by the protein synthesis inhibitor studies (125 – 127). Furthermore they support the view that it is the translation of the histone mRNA itself rather than the synthesis of a *trans*-acting factor that is important for histone mRNA destabilization.

9. Transcriptional termination and gene regulation

The importance of transcriptional termination to the regulation of prokaryotic gene expression is well documented. The mechanisms of attenuation and anti-termination are of profound significance to bacteria (1). However, these mechanisms are dependent on the tight coupling of transcription and translation, made possible by the fact that bacteria do not separate their chromosome from the cytoplasm. The possession of a nucleus in eukaryotes must significantly alter the role termination can have in gene regulation. Indeed, as discussed in Section 6.3, it is formally possible that, since mRNA 3' termini are formed by RNA processing, transcriptional termination is simply a consequence of the 3' processing event. However, as documented in this section, the fact that transcriptional termination can regulate gene expression by premature termination or pausing within a gene or by transcriptional interference between genes makes it clear that transcriptional termination can be a sequence-specific and regulated process.

9.1 Regulation of gene expression by premature termination or transcriptional pausing

9.1.1 Adenovirus

Transcription reads continuously through most of the adenovirus genome (129). Different mRNAs are produced from this single primary transcript, about 25 kb long, both by differential poly(A) site selection (130) as well as by differential splicing (see ref. 131 for review). However at early times of infection a different termination site appears to be active, less than half way through the viral genome (132). This effectively restricts mRNA production to only the first two poly(A) sites. The mechanism controlling this termination and presumably anti-termination process is unknown.

9.1.2 Polyoma and SV40

In the small circular DNA tumor viruses, polyoma and SV40, the role of transcriptional termination is unclear. For polyoma, large primary RNA transcripts have been identified that must be multiple concatameric transcripts of the whole viral genome (133,134). In other words, no efficient termination process appears to operate. Interestingly only a small fraction of these concatameric transcripts are processed to polyadenylated mRNA, which may suggest a linkage between poor termination and poor 3' end processing (134) (see Section 6.3). For SV40, efficient transcription occurs at least 1 kb past the late mRNA poly(A) site (135). Furthermore, a pause site has been identified soon after the late promoter (136,137). This pause site may have a role in the regulation of SV40 transcription although its physiological importance remains to be established.

9.1.3 Immunoglobulin μ and δ constant region genes

In early B cells the μ constant region is exclusively utilized although the earliest B cell progenitors synthesize the membrane bound IgM antibody while slightly later B cells synthesize secreted IgM. The differential expression of these two antibody forms is regulated by different poly(A) site utilization as described in Section 8.3. The RNA splicing pattern for the production of IgM by these B cells is shown beneath the gene map in *Figure 7*. The next stage in B cell development results in the co-expression of both μ and δ constant regions making both IgM and IgD antibodies. The molecular basis of this altered pattern of expression is caused by the differential splicing out of all the μ constant region exons so as to join the VDJ segment to the δ constant region exons (138) as shown above the gene map in *Figure 7*. Surprisingly a third level of transcriptional regulation also exists. It has been demonstrated by

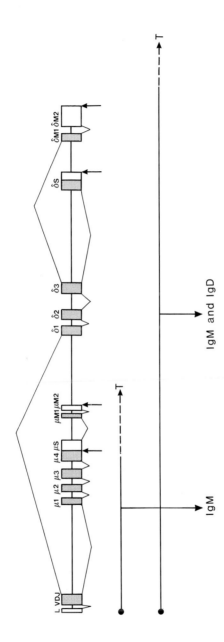

Figure 7. Differential termination in the transcription of immunoglobulin μ and δ constant region genes. T denotes termination; the dashed lines indicate uncertainty in the precise location of transcription termination. L denotes leader exon. V denotes variable, D denotes diversity and J denotes joining portions of the VDJ exon. $\mu 1 - \mu 4$ denote the four μ constant region exons. μS, $\mu M1$ and $\mu M2$ denote the secreted and membrane μ exons. $\delta 1 - 3$ denote three δ constant region exons. δS, $\delta M1$ and $\delta M2$ denote the C-terminal exons of the δ constant region for the secreted and membrane forms, respectively. Open boxes denote 5' and 3' non-coding region exons. Hatched boxes denote coding sequence exons. Horizontal lines denote introns. Lines joining exons above and below the gene map indicate splicing reactions.

nuclear run-off analysis that in B cells synthesizing only IgM antibody, termination of transcription occurs between the μ and δ constant region exons (*Figure 7*). Alternatively, in B cells that make both IgM and IgD by differential splicing the termination site between μ and δ is lost and transcription proceeds past the δ constant region exons (*Figure 7*). These experiments provide a clear example of differential termination of transcription being used as a means of controlling gene expression (139).

9.1.4 Mouse β-globin gene

It has been reported that under some circumstances transcription terminates or pauses in the first intron of the mouse β-globin gene (76,140). However, following induction of the gene using DMSO treatment of β-globin expressing Friend cells, this pause site disappears and transcription proceeds to the regular termination site approximately 500 bp past the gene's poly(A) site.

9.1.5 Human c-myc gene

The c-*myc* gene has been extensively studied due to the association of various types of cancer with a mutated c-*myc* gene (141). Interestingly this gene also appears to utilize transcriptional pausing in the regulation of its expression (108,142). Nuclear run-off analysis indicates that nascent transcription pauses at the end of the first exon. Furthermore this pause site is especially predominant in cell lines which express reduced levels of c-*myc* following induced differentiation.

The several different examples of transcription termination or pausing within a gene documented in this section clearly suggest the importance of this mechanism to the control of gene expression.

9.2 Transcriptional interference and termination

The phenomenon of transcriptional interference, or promoter occlusion as it was first called, was originally discovered in bacteria. Following the integration of bacteriophage λ into the *Escherichia coli* genome it was observed that the *gal* operon promoter P91 close to the site of λ integration was severely inhibited by transcription off the λP_L promoter (143). More recently it has been shown that this same phenomenon of transcriptional interference plays a very important role in the mechanism of insertional oncogenesis by the avian leukosis retrovirus (ALV). Retroviruses contain two identical copies of a strong enhancer-stimulated promoter, one in each long terminal repeat (LTR) sequence. It has been demonstrated that this insertional oncogenesis is often caused by the integration of ALV into a host genome adjacent to the c-*myc* gene and that transcription initiates on the 3′ LTR promoter reading into the c-*myc* gene. Interestingly, active 3′ LTR transcription and hence *myc*-mediated transformation only occur in viral integrants that have a damaged or deleted 5′ LTR sequence

(144,145). These observations prompted experiments that directly test the hypothesis that interference occurs between the two LTR promoters (146). It was shown that the 3' LTR promoter was only functional when either the 5' LTR promoter was deleted or a DNA fragment containing an SV40 poly(A) site, presumably also including a linked transcriptional termination process, was added between the two LTR sequences.

We have extended these observations to the possibility that separate but closely-linked genes might similarly interfere with each other. We demonstrated that when two human α-globin genes were artificially placed in tandem, transcription off the second (3' positioned) gene was inhibited by transcription off the first (5' positioned) gene. Furthermore this interference effect could be alleviated by placing termination signals between two α-globin genes (147). These data have direct implications for the role of transcriptional termination in the expression of RNA

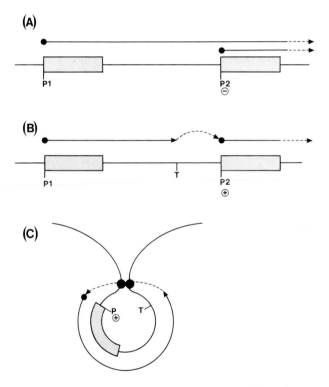

Figure 8. Hypothetical mechanisms for the interaction of transcriptional termination and initiation in two adjacent genes (**A** and **B**) and within one gene (**C**). Hatched boxes denote the gene sequences. Connecting lines denote intergenic sequence. P denotes promoter and T denotes termination. The arrows denote passage of RNA polymerase during transcription, while dotted arrows indicate possible migration of RNA polymerase from termination site to adjacent promoter. ⊕ activated promoter. ⊖ inhibited promoter.

polymerase II genes. Since all genes must have neighboring genes on a particular chromosome, interference would be expected to occur between them if no termination of transcription occurred. These considerations lead to an interesting model for the use of transcriptional interference and termination in the control of gene expression (147,148), as shown in *Figure 8A*. Essentially, regulation of the first gene will indirectly and inversely regulate the activity of the second gene. In other words, the more active the first gene is the more it will interfere with the 3′ positioned second gene. Furthermore the selective use of termination signals between the two genes might provide a means of regulating the 3′ positioned gene. Evidence is now accumulating for different examples of transcription interference (149,150), although so far no natural application of this regulatory mechanism has been identified.

Transcriptional interference may be caused by two quite different mechanisms, one through inhibition and the other through activation of the second gene. In the first case, as already described, initiation of transcription off the second gene could be impeded by elongating polymerase molecules reading through from the first gene. Termination would alleviate this inhibition (*Figure 8A*). Alternatively, termination of transcription after one gene could increase transcription off the second gene by releasing the polymerase molecule from the DNA template close to the second gene, effectively increasing the availability of polymerase for the promoter of the second gene (*Figure 8B*). This second mechanism is very similar to the enhancement of transcription observed in Pol I genes by the possession of a terminator sequence close to the major Pol I promoter (see Section 7.3). This model for the activation of genes in tandem by releasing RNA polymerase at termination sites to make the polymerase available for the promoter of the adjacent gene can be adapted to a single gene context. It is well documented that the DNA of the eukaryotic chromosomes may exist in looped conformations (151,152). If such a DNA loop contained a single gene, then termination of transcription at the end of the gene could release the polymerase to re-initiate transcription on the promoter of the same gene (see *Figure 8C*). The roles of transcriptional termination in the prevention of transcription interference or in the enhancement of initiation of transcription are intriguing possibilities for the regulation of eukaryotic genes. It remains to be established whether or not these mechanisms are important to eukaryotic gene expression.

10. References

1. Platt,T. (1986) Transcription termination and the regulation of gene expression. *Annu. Rev. Biochem.*, **55**, 339.
2. Brennan,C.A., Dombroski,A.J. and Platt,T. (1987) Transcription termination factor Rho is an RNA-DNA helicase. *Cell,* **48**, 945.
3. Lim,L. and Canellakis,E.S. (1970) Adenine-rich polymer associated with rabbit reticulocyte messenger RNA. *Nature,* **227**, 710.

4. Edmonds,M., Vaughan,M.H. and Nakazato,H. (1971) Polyadenylic acid sequences in heterogeneous nuclear RNA and rapidly labeled polyribosomal RNA of HeLa cells. Possible evidence for a precursor relationship. *Proc. Natl. Acad. Sci. USA,* **68,** 1136.

5. Lee,S.V., Mendecki,J. and Brawerman,G. (1971) A polynucleotide segment rich in adenylic acid in the rapidly-labeled polyribosomal RNA component of mouse sarcoma 180 ascites cells. *Proc. Natl. Acad. Sci. USA,* **68,** 1331.

6. Birnboim,H.C., Mitchel,R.E.J. and Straus,N.A. (1973) Analysis of long pyrimidine polynucleotides in HeLa cell nucler DNA: absence of polydeoxythymidylate. *Proc. Natl. Acad. Sci. USA,* **70,** 2189.

7. Edmonds,M. and Winters,M.A. (1976) Polyadenylate polymerases. *Prog. Nucleic Acids Res. Mol. Biol.,* **17,** 149.

8. Aviv,H. and Leder,P. (1972) Purification of biologically active globin messenger RNA by chromatography on oligothymidylic acid-cellulose. *Proc. Natl. Acad. Sci. USA,* **69,** 1408.

9. Efstratiadis,A., Kafatos,F.C., Maxam,A.M. and Maniatis,T. (1976) Enzymatic *in vitro* synthesis of globin genes. *Cell,* **7,** 279.

10. Huez,G., Marbaix,G., Hubert,E., Leclerq,M., Nadel,U., Soreq,H., Solomon,R., Lebleu,B., Revel,M. and Littauer,U.Z. (1974) Role of the polyadenylate segment in the translation of globin messenger RNA in *Xenopus* oocytes. *Proc. Natl. Acad. Sci. USA,* **71,** 3143.

11. Drummond,D.R., Armstrong,J. and Colman,A. (1985) The effect of capping and polyadenylation on the stability, movement and translation of synthetic messenger RNAs in *Xenopus* oocytes. *Virology,* **13,** 7375.

12. Proudfoot,N.J. and Brownlee,G.G. (1976) The 3' non-coding region sequences in eukaryotic messenger RNA. *Nature,* **263,** 211.

13. Nevins,J.R. (1983) The pathways of eukaryotic mRNA formation. *Annu. Rev. Biochem.,* **52,** 441.

14. Fitzgerald,M. and Shenk,T. (1981) The sequence of 5'-AAUAAA-3' forms part of the recognition site for polyadenylation of late SV40 mRNAs. *Cell,* **24,** 251.

15. Montell,C., Fisher,E.F., Caruthers,M.H. and Berk,A.J. (1983) Inhibition of RNA cleavage but not polyadenylation by a point mutation in mRNA 3' consensus sequence AAUAAA. *Nature,* **305,** 600.

16. Higgs,D.R., Goodbourn,S.E.Y., Lamb,J., Clegg,J.B., Weatherall,D.J. and Proudfoot, N.J. (1983) α-Thalassaemia caused by a polyadenylation signal mutation. *Nature,* **306,** 398.

17. Orkin,S.H., Cheng,T.-C., Antonarakis,S.E. and Kazaian,H.H., Jr (1985) Thalassemia due to a mutation in the cleavage-polyadenylation signal of the human β-globin gene. *EMBO J.,* **4,** 453.

18. Whitelaw,E. and Proudfoot,N.J. (1986) α-Thalassaemia caused by a poly(A) site mutation reveals that transcriptional termination is linked to 3' end processing in the human α2 globin gene. *EMBO J.,* **5,** 2915.

19. Wickens,M. and Stephenson,P. (1984) Role of the conserved AAUAAA sequence: four AAUAAA point mutants prevent messenger RNA 3' end formation. *Science,* **226,** 1045.

20. Setzer,D.R., McGrogan,M. and Schimke,R.T. (1982) Nucleotide sequence surrounding multiple polyadenylation sites in the mouse dihydrofolate reductase gene. *J. Biol. Chem.,* **257,** 5143.

21. Tosi,M., Young,R.A., Hagenbuchle,O. and Schibler,U. (1981) Multiple polyadenylation sites in a mouse α amylase gene. *Nucleic Acids Res.,* **9,** 2313.

22. Aho,S., Tate,V. and Boedtker,H. (1983) Multiple 3' ends of the chicken pro α2(I) collagen gene. *Nucleic Acids Res.,* **11,** 5443.

23. Simonsen,C.C. and Levinson,A.D. (1983) Analysis of processing and polyadenylation signals of the hepatitis B virus surface antigen gene by using simian virus 40-hepatitis B virus chimeric plasmids. *Mol. Cell. Biol.,* **3,** 2250.

24. McDevitt,M.A., Imperiale,M., Ali,H. and Nevins,J.R. (1984) Requirement of a downstream sequence for generation of a poly(A) addition site. *Cell,* **37,** 993.

25. Gil,A. and Proudfoot,N.J. (1984) A sequence downstream of AAUAAA is required for rabbit β-globin mRNA 3' end formation. *Nature,* **312,** 473.

26. Woychik,R.P., Lyons,R.H., Post,L. and Rottmann,F.M. (1984) Requirement for the 3' flanking region of the bovine growth hormone for accurate polyadenylation. *Proc. Natl. Acad. Sci. USA,* **81,** 3944.

27. Sadofsky,M. and Alwine,J.C. (1984) Sequences on the 3' side of hexanucleotide AAUAAA affect efficiency of cleavage at the polyadenylation site. *Mol. Cell. Biol.*, **4**, 1460.
28. Hart,R.P., McDevitt,M.A., Ali,H. and Nevins,J.R. (1985) Definition of essential sequences and functional equivalence of elements downstream of the adenovirus E2A and the early simian virus 40 polyadenylation sites. *Mol. Cell. Biol.*, **5**, 2975.
29. Cole,C.N. and Stacey,T.P. (1985) Identification of sequences in the herpes simplex virus type 1 thymidine kinase gene required for efficient processing and polyadenylation. *Mol. Cell. Biol.*, **3**, 267.
30. McLauchlan,J., Gaffney,D., Whitton,J.L. and Clements,J.B. (1985) The consensus sequence YGTGTTYY located downstream from the AATAAA signal is required for efficient formation of mRNA 3' termini. *Nucleic Acids Res.*, **13**, 1347.
31. McDevitt,M.A., Hart,R.P., Wong,W.W. and Nevins,J.R. (1986) Sequences capable of restoring poly(A) site function define two distinct downstream elements. *EMBO J.*, **5**, 2907.
32. Gil,A. and Proudfoot,N.J. (1987) Position-dependent sequence elements downstream of AAUAAA are required for efficient rabbit β-globin mRNA 3' end formation. *Cell*, **49**, 399.
33. Zhang,F., Denome,R.M. and Cole,C.N. (1986) Fine-structure analysis of the processing and polyadenylation region of the herpes simplex virus type 1 thymidine kinase gene by using linker scanning, internal deletion and insertion mutations. *Mol. Cell. Biol.*, **6**, 4611.
34. Mason,P.J., Elkington,J.A., Malgorzata,M.L., Jones,M.B. and Williams,J.G. (1986) Mutations downstream of the polyadenylation site of a *Xenopus* β-globin mRNA affect the position but not the efficiency of 3' processing. *Cell*, **46**, 263.
35. Manley,J.L. (1983) Accurate and specific polyadenylation of mRNA precursors in a soluble whole-cell lysate. *Cell*, **33**, 595.
36. Moore,C.L. and Sharp,P.A. (1984) Site-specific polyadenylation in a cell-free reaction. *Cell*, **36**, 581.
37. Moore,C.L. and Sharp,P.A. (1985) Accurate cleavage and polyadenylation of exogenous RNA substrate. *Cell*, **41**, 485.
38. Melton,D.A., Krieg,P.A., Rebagliati,M.R., Maniatis,T., Zinn,K. and Green,M.R. (1984) Efficient *in vitro* synthesis of biologically active RNA and RNA hybridization probes from plasmids containing a bacteriophage SP6 promoter. *Nucleic Acids Res.*, **12**, 7035.
39. Zarkower,D., Stephenson,P., Sheets,M. and Wickens,M. (1986) The AAUAAA sequence is required both for cleavage and for polyadenylation of simian virus 40 pre-mRNA *in vitro*. *Mol. Cell. Biol.*, **6**, 2317.
40. Hart,R.P., McDevitt,M.A. and Nevins,J.R. (1985) Poly(A) site cleavage in a HeLa nuclear extract is dependent on downstream sequences. *Cell*, **43**, 677.
41. Sperry,A.D. and Berget,S.M. (1986) *In vitro* cleavage of the simian virus 40 early polyadenylation site adjacent to a required downstream TG sequence. *Mol. Cell. Biol.*, **6**, 4734.
42. Berget,S.M. (1984) Are U4 small nuclear ribonucleoproteins involved in polyadenylation? *Nature*, **309**, 179.
43. Hashimoto,C. and Steitz,J.A. (1986) A small nuclear ribonucleoprotein associates with the AAUAAA polyadenylation signal *in vitro*. *Cell*, **45**, 581.
44. Berget,S.M. and Robberson,B.L. (1986) U1, U2 and U4/U6 small nuclear ribonucleoproteins are required for *in vitro* splicing but not polyadenylation. *Cell*, **46**, 691.
45. Maniatis,T. and Reed,R. (1987) The role of small nuclear ribonucleoprotein particles in pre-mRNA splicing. *Nature*, **325**, 673.
46. Reddy,R. and Busch,H. (1981) U SnRNAs of nuclear SnRNPs. In *The Cell Nucleus*. Bush,H. (ed.), Academic Press, New York, Vol. 8, p. 261.
47. Hentschel,C.C. and Birnstiel,M.L. (1981) The organization and expression of histone gene families. *Cell*, **25**, 301.
48. Old,R.W. and Woodland,H.R. (1984) Histone genes: not so simple after all. *Cell*, **38**, 624.
49. Birnstiel,M.L., Busslinger,M. and Strub,K. (1985) Transcription termination and 3' processing: the end is in site! *Cell*, **41**, 349.
50. Birchmeier,C., Grosschedl,R. and Birnstiel,M. (1982) Generation of authentic 3' termini of an H2A mRNA *in vivo* is dependent on a short inverted DNA repeat and on spacer sequences. *Cell*, **28**, 739.

51. Birchmeier,C., Folk,W. and Birnstiel,M.L. (1983) The terminal RNA stem-loop structure and 80 bp of spacer DNA are required for the formation of 3' termini of sea urchin H2A mRNA. *Cell,* **35,** 433.
52. Georgiev,O. and Birnstiel,M.L. (1985) The conserved CAAGAAAGA spacer sequence is an essential element for the formation of 3' termini of the sea urchin H3 histone mRNA by RNA processing. *EMBO J.,* **4,** 481.
53. Stunnenburg,H.G. and Birnstiel,M.L. (1982) Bioassay for components regulating eukaryotic gene expression: a chromosomal factor involved in the generation of histone mRNA 3' termini. *Proc. Natl. Acad. Sci. USA,* **79,** 6201.
54. Galli,G., Hofstetter,H., Stunnenberg,H.G. and Birnstiel,M.L. (1983) Biochemical complementation with RNA in the *Xenopus* oocyte. A small RNA is required for the generation of 3' histone mRNA termini. *Cell,* **34,** 823.
55. Price,D.H. and Parker,C.S. (1984) The 3' end of *Drosophila* histone H3 mRNA is produced by a processing activity *in vitro. Cell,* **38,** 423.
56. Birchmeier,C., Schumperli,D., Sconzo,G. and Birnstiel,M.L. (1984) 3' editing of mRNAs: sequence requirements and involvement of a 60-nucleotide RNA in maturation of histone mRNA precursors. *Proc. Natl. Acad. Sci. USA,* **81,** 1057.
57. Krieg,P.A. and Melton,D.A. (1984) Formation of the 3' end of histone mRNA by post-transcriptional processing. *Nature,* **308,** 203.
58. Gick,O., Kramer,A., Keller,W. and Birnstiel,M.L. (1986) Generation of histone mRNA 3' ends by endonucleolytic cleavage of the pre-mRNA in a SnRNP-dependent *in vitro* reaction. *EMBO J.,* **5,** 1319.
59. Strub,K., Galli,G., Busslinger,M. and Birnstiel,M.L. (1984) cDNA sequences of the sea urchin U7 small nuclear RNA suggest specific contacts between histone mRNA precursor and U7 RNA during RNA processing. *EMBO J.,* **3,** 2801.
60. Strub,K. and Birnstiel,M.L. (1986) Genetic complementation in the *Xenopus* oocyte: co-expression of sea urchin histone and U7 RNAs restores 3' processing of H3 pre-mRNA in the oocyte. *EMBO J.,* **5,** 1675.
61. Schaufele,F., Gilmartin,G.M., Bannwarth,W. and Birnstiel,M.L. (1986) Compensatory mutations suggest that base-pairing with a small nuclear RNA is required to form the 3' end of H3 messenger RNA. *Nature,* **323,** 777.
62. Murphy,J.T., Burgess,R.R., Dahlberg,J.E. and Lund,E. (1982) Transcription of a gene for human U1 small nuclear RNA. *Cell,* **29,** 265.
63. Krol,A., Carbon,P., Ebel,J.-P. and Appel,B. (1987) *Xenopus tropicalis* U6 snRNA genes transcribed by polIII contain the upstream promoter elements used by Pol II dependent U snRNA genes. *Nucleic Acids Res.,* **15,** 2463.
64. Hernandez,N. (1985) Formation of the 3' end of U1 snRNA is directed by a conserved sequence located downstream of the coding sequence. *EMBO J.,* **4,** 1827.
65. Yuo,C., Ares,M.,Jr and Weiner,A.M. (1985) Sequences required for 3' end formation of human U2 small nuclear RNA. *Cell,* **42,** 193.
66. Ciliberto,G., Dathan,N., Frank,R., Philipson,L. and Mattaj,I.W. (1986) Formation of the 3' end on U snRNAs requires at least three sequence elements. *EMBO J.,* **5,** 2931.
67. Kunkel,G.R. and Pederson,T. (1985) Transcriptional boundaries of U1 small nuclear RNA. *Mol. Cell. Biol.,* **5,** 2332.
68. Hernandez,N. and Weiner,A.M. (1986) Formation of the 3' end of U1 snRNA requires compatible snRNA promoter elements. *Cell,* **47,** 249.
69. Neuman de Vegvar,H.E., Lund,E. and Dahlberg,J.E. (1986) 3' end formation of U1 snRNA precursors is coupled to transcription from snRNA promoters. *Cell,* **47,** 259.
70. Zaret,K.S. and Sherman,F. (1982) DNA sequence required for efficient transcription termination in yeast. *Cell,* **28,** 563.
71. Henikoff,S., Kelly,J.D. and Cohen,E.H. (1983) Transcription terminates in yeast distal to a control sequence. *Cell,* **33,** 607.
72. Zaret,K.S. and Sherman,F. (1984) Mutationally altered 3' ends of yeast CYC1 mRNA affect transcript stability and translational efficiency. *J. Mol. Biol.,* **176,** 107.
73. Toole,J.J., Knopf,J.L., Wosney,J.M., Sultzman,L.A., Buecker,J.L., Pittman,D.D., Kaufman,R.J., Brown,E., Shoemaker,C., Orr,E.C., Amphlett,G.W., Foster,B.W., Cou,M.L., Knutson,G.J., Fass,D.N. and Hewick,R.M. (1984) Molecular cloning of a cDNA encoding human antihaemophilic factor. *Nature,* **312,** 342.
74. Anson,D.S., Choo,K.H., Rees,D.J.G., Giannelli,F., Gould,K., Huddleston,J.A. and Brownlee,G.G. (1984) Gene structure of human anti-haemophilic factor IX. *EMBO J.,* **3,** 1053.

75. Groudine,M., Peretz,M. and Weintraub,H. (1981) Transcriptional regulation of haemoglobin switching in chicken embryos. *Mol. Cell. Biol.,* **1**, 281.
76. Hofer,E., Hofer-Warbinek,R. and Darnell,J.E.,Jr (1982) Globin RNA transcription: a possible termination site and demonstration of transcription control correlated with altered chromatin structure. *Cell,* **29**, 887.
77. Citron,B., Falck-Pederson,E., Salditt-Georgieff,M. and Darnell,J.E.,Jr (1984) Transcription termination occurs within a 1000 bp region downstream from the poly(A) site of the mouse β-globin (major) gene. *Nucleic Acids Res.,* **12**, 8723.
78. Hagenbuchle,O., Wellauer,P.K., Cribbs,D.L. and Schibler,U. (1984) Termination of transcription in the mouse α-amylase gene Amy-2^a occurs at multiple sites downstream of the polyadenylation site. *Cell,* **38**, 737.
79. LeMeur,M.A., Galliot,B. and Gerlinger,P. (1984) Termination of the ovalbumin gene transcription. *EMBO J.,* **3**, 2779.
80. Rohrbaugh,M.L., Johnson,J.E., III, James,M.D. and Hardison,R.C. (1985) Transcription unit of the rabbit β1 globin gene. *Mol. Cell. Biol.,* **5**, 147.
81. Weintraub,H., Larsen,A. and Groudine,M. (1981) α-Globin gene switching during the development of chicken embryos: expression and chromosome strucutre. *Cell,* **24**, 333.
82. Sheffery,M., Marks,P.A. and Rifkind,R.A. (1984) Gene expression in murine erythroleukemia cells: transcriptional control and chromatin structure of the α1 globin gene. *J. Mol. Biol.,* **172**, 417.
83. Frayne,E.G., Leys,E.J., Crouse,G.F., Hook,A.G. and Kellems,R.E. (1984) Transcription of the mouse dihydrofolate reductase gene proceeds unabated through seven polyadenylation sites and terminates near a region of repeated DNA. *Mol. Cell. Biol.,* **4**, 2921.
84. Frayne,E.G. and Kellems,R.E. (1986) Structural features of the murine dihydrofolate reductase transcription termination region: identification of a conserved DNA sequence element. *Nucleic Acids Res.,* **14**, 4113.
85. Amara,S.G., Evans,R.M. and Rosenfeld,M.G. (1984) Calcitonin/calcitonin gene-related peptide transcription unit: tissue-specific expression involves selective use of alternative polyadenylation sites. *Mol. Cell. Biol.,* **4**, 2151.
86. Falck-Pedersen,E., Logan,J., Shenk,T. and Darnell,J.E.,Jr (1985) Transcription termination within the E1A gene of adenovirus induced by insertion of the mouse β-major globin terminator element. *Cell,* **40**, 897.
87. Sato,K., Ito,R., Baek,K.-H. and Agarwal,K. (1986) A specific DNA sequence controls termination in the gastrin gene. *Mol. Cell. Biol.,* **6**, 1032.
88. Baek,K.-H., Sato,K., Ito,R. and Agarwal,K. (1986) RNA polymerase II transcription terminates at a specific DNA sequence in a HeLa cell-free reaction. *Proc. Natl. Acad. Sci. USA,* **83**, 7623.
89. Johnson,M.R., Norman,C., Reeve,M.A., Scully,J. and Proudfoot,N.J. (1986) Tripartite sequences within and 3′ to the sea urchin H2A histone gene display properties associated with a transcriptional termination process. *Mol. Cell. Biol.,* **6**, 4008.
90. Cozzarelli,N.R., Gerrard,S.P., Schlissel,M., Brown,D.D. and Boenhagen,D.F. (1983) Purified RNA polymerase III accurately and efficiently terminates transcription of 5S RNA genes. *Cell,* **34**, 829.
91. Bogenhagen,D.F. and Brown,D.D. (1981) Nucleotide sequences in *Xenopus* 5S DNA required for transcription termination. *Cell,* **24**, 261.
92. Clarkson,S.G. (1983) Transfer RNA genes. In *Eukaryotic Genes, Their Structure, Activity and Regulation.* Maclean,N., Gregory,S.P. and Flavell,R.A. (eds), Butterworths, Chapter 14.
93. Lewin,B. (1985) *Genes II.* Wiley, Chichester, Chapters 25 and 26.
94. Sollner-Webb,B. and Tower,J. (1986) Transcription of cloned eukaryotic ribosomal genes. *Annu. Rev. Biochem.,* **55**, 801.
95. Reeder,R.H. (1984) Enhancers and ribosomal gene spacers. *Cell,* **38**, 349.
96. Miller,O.L. and Beatty,B.R. (1969) Visualisation of nucleolar genes. *Science,* **164**, 955.
97. Grummt,I., Maier,U., Ohrlein,A., Hassouna,N. and Bachellerie,J.-P. (1985) Transcription of mouse rDNA terminates downstream of the 3′ end of 28S RNA and involves interaction of factors with repeated sequences in the 3′ spacer. *Cell,* **43**, 801.
98. Grummt,I., Rosenbauer,H., Nedermeyer,I., Maier,U. and Ohrlein,A. (1986) A repeated 18 bp sequence motif in the mouse rDNA spacer mediates binding of a nuclear factor and transcription termination. *Cell,* **45**, 837.

99. Moss,T. (1983) A transcriptional function for the repetitive ribosomal spacer of *Xenopus laevis. Nature,* **302**, 223.

100. Tautz,D. and Dover,G.A. (1986) Transcription of the tandem array of ribosomal DNA in *Drosophila melanogaster* does not terminate at any fixed point. *EMBO J.,* **5**, 1267.

101. Labhart,P. and Reeder,R.H. (1986) Characterisation of three sites of RNA 3' end formation in the *Xenopus* ribosomal gene spacer. *Cell,* **45**, 431.

102. McStay,B. and Reeder,R.H. (1986) A termination site for *Xenopus* RNA polymerase I also acts as an element of an adjacent promoter. *Cell,* **47**, 913.

103. Grummt,I., Kuhn,A., Bartsch,I. and Rosenbauer,H. (1986) A transcription terminator located upstream of the mouse rDNA initiation site affects rRNA synthesis. *Cell,* **47**, 901.

104. Henderson,S. and Sollner-Webb,B. (1986) A transcriptional terminator is a novel element of the promoter of the mouse ribosomal RNA gene. *Cell,* **47**, 891.

105. Newbury,S.F., Smith,N.H., Robinson,E.C., Hiles,I.D. and Higgins,C.F. (1987) Stabilization of translationally active mRNA by prokaryotic REP sequences. *Cell,* **48**, 297.

106. Shaw,G and Kamen,R. (1986) A conserved AU sequence from the 3' untranslated region of GM-CSF mRNA mediates selective mRNA degradation. *Cell,* **46**, 659.

107. Littauer,U.Z. and Soreq,H. (1982) The regulatory function of poly(A) and adjacent 3' sequences in translated RNA. *Prog. Nucleic Acid Res.,* **27**, 53.

108. Nepveu,A. and Marcu,K.B. (1986) Intragenic pausing and anti-sense transcription within the murine c-*myc* locus. *EMBO J.,* **5**, 2859.

109. Leff,S.E., Rosenfeld,M.G. and Evans,R.M. (1986) Complex transcriptional units: diversity in gene expression by alternative RNA processing. *Annu. Rev. Biochem.,* **55**, 1091.

110. Perricaudet,M., LeMoullec,J.M., Tiollais,P. and Petterson,U. (1980) Structure of two adenovirus type 12 transforming polypeptides and their evolutionary implications. *Nature,* **288**, 174.

111. Parnes,J.R., Robinson,J.R. and Seidman,J.G. (1983) Multiple mRNA species with distinct 3' termini are transcribed from the β_2-microglobulin gene. *Nature,* **302**, 449.

112. Capetanaki,Y.G., Ngai,J., Flytzanis,C.N. and Lazarides,E. (1983) Tissue-specific expression of two mRNA species transcribed from a single vimentin gene. *Cell,* **35**, 411.

113. Alt,F.W., Bothwell,A.L.M., Knapp,M., Siden,E., Mather,E., Koshland,M. and Baltimore,D. (1980) Synthesis of secreted and membrane-bound Mu heavy chains is directed by mRNA that differ at their 3' ends. *Cell,* **20**, 293.

114. Roger,J., Early,P., Carter,C., Calame,K., Bond,M., Hood,L. and Wall,R. (1980) Two mRNAs with different 3' ends encode membrane-bound and secreted forms of immunoglobulin μ chain. *Cell,* **20**, 303.

115. Early,P., Rogers,J., Davis,M., Calame,K., Bond,M., Wall,R. and Hood,L. (1980) Two mRNAs can be produced from a single immunoglobulin μ gene by alternative RNA processing pathways. *Cell,* **20**, 313.

116. Amara,S.G., Jones,V., Rosenfeld,M.G., Ong,E.S. and Evans,R.M. (1982) Alternative RNA processing in calcitonin gene expression generates mRNAs encoding different polypeptide products. *Nature,* **298**, 240.

117. Yuan,D. and Tucker,P.W. (1984) Transcriptional regulation of the $\mu - \delta$ heavy chain locus in normal murine B lymphocytes. *J. Exp. Med.,* **160**, 564.

118. Kemp,D.J., Morahan,G., Cowman,A.F. and Harris,A.W. (1983) Production of RNA for secreted immunoglobulin μ chains does not require transcriptional termination 5' to the μM exons. *Nature,* **301**, 84.

119. Danner,D. and Leder,P. (1985) Role of an RNA cleavage/poly(A) addition site in the production of membrane-bound and secreted IgM mRNA. *Proc. Natl. Acad. Sci. USA,* **82**, 8658.

120. Leff,S.E., Evans,R.M. and Rosenfeld,M.G. (1987) Splice commitment dictates neuron-specific alternative RNA processing in calcitonin/CGRP gene expression. *Cell,* **48**, 517.

121. Heintz,N., Sive,H.L. and Roeder,R.G. (1983) Regulation of human histone gene expression: kinetics of accumulation and changes in the rate of synthesis and in the half-lives of individual histone mRNAs during the HeLa cell cycle. *Mol. Cell. Biol.,* **3**, 539.

122. Hanly,S.M., Bleecker,G.C. and Heintz,N. (1985) Identification of promoter elements

necessary for transcriptional regulation of a human histone H4 gene *in vitro*. *Mol. Cell. Biol.,* **5**, 380.

123. Luscher,B., Stauber,C., Schindler,R. and Schumperli,D. (1985) Faithful cell cycle regulation of a recombinant mouse histone H4 gene is controlled by sequences in the 3' terminal part of the gene. *Proc. Natl. Acad. Sci. USA,* **82**, 4389.

124. Stauber,C., Luscher,B., Eckner,R., Lotscher,E. and Schumperli,D. (1986) A signal regulating mouse histone H4 mRNA levels in a mammalian cell cycle mutant and sequences controlling RNA 3' processing are both contained within the same 80 bp fragment. *EMBO J.,* **5**, 3297.

125. Butler,W.G. and Mueller,G.C. (1973) Control of histone synthesis in HeLa cells. *Biochim. Biophys. Acta,* **294**, 481.

126. Stahl,H. and Gallwitz,J. (1977) Fate of histone messenger RNA in synchronised HeLa cells in the absence of initiation of protein synthesis. *Eur. J. Biochem.,* **72**, 385.

127. Stimac,E., Groppi,V.E.,Jr and Coffino,P. (1984) Inhibition of protein synthesis stabilizes histone mRNA. *Mol. Cell. Biol.,* **4**, 2082.

128. Graves,R.A., Pandey,N.B., Chodchoy,N. and Marzluff,W.F. (1987) Translation is required for regulation of histone mRNA degradation. *Cell,* **48**, 615.

129. Fraser,N.W., Nevins,J.R., Ziff,E. and Darnell,J.E. (1979) The major late adenovirus type 2 transcription unit: termination is downstream from the last poly(A) site. *J. Mol. Biol.,* **129**, 643.

130. Fraser,N. and Ziff,E. (1978) RNA structures near poly(A) of adenovirus-2 late messenger RNAs. *J. Mol. Biol.,* **124**, 27.

131. Ziff,E.B. (1980) Transcription and RNA processing by the DNA tumour viruses. Nature, **287**, 491.

132. Nevins,J.R. and Wilson,M.C. (1981) Regulation of adenovirus-2 gene expression at the level of transcriptional termination and RNA processing. *Nature,* **290**, 113.

133. Treisman,R. and Kamen,R. (1981) Structure of polyomer virus late nuclear RNA. *J. Mol. Biol.,* **148**, 273.

134. Acheson,N.H. (1984) Kinetics and efficiency of polyadenylation of late polyomavirus nuclear RNA: generation of oligomeric polyadenylated RNAs and their processing into mRNA. *Mol. Cell. Biol.,* **4**, 722.

135. Ford,J. and Hsu,M.-T. (1978) Transcription pattern of *in vivo* labelled late SV40 RNA: equimolar transcription beyond mRNA 3' terminus. *J. Virol.,* **28**, 795.

136. Hay,N., Skolnick-David,H. and Aloni,Y. (1982) Attenuation in the control of SV40 gene expression. *Cell,* **29**, 183.

137. Hay,N. and Aloni,Y. (1984) Attentuation in SV40 as a mechanism of transcription termination by RNA polymerase B. *Nucleic Acids Res.,* **12**, 1401.

138. Blattner,F.R. and Tucker,P.W. (1984) The molecular biology of immunoglobulin D. *Nature,* **307**, 417.

139. Mather,E.L., Nelson,K.J., Haimovich,J. and Perry,R.P. (1984) Mode of regulation of immunoglobulin μ and δ chain expression varies during B-lymphocyte maturation. *Cell,* **36**, 329.

140. Sheffrey,M., Rifkind,R.A. and Marks,P.A. (1982) Murine erythroleukemia cells differentiation: DNaseI hypersensitivity and DNA methylation near the globin genes. *Proc. Natl. Acad. Sci. USA,* **79**, 1620.

141. Bishop,J.M. (1985) Viral oncogenes. *Cell,* **42**, 23.

142. Bentley,D.L. and Groudine,M. (1986) A block to elongation is largely responsible for decreased transcription of c-*myc* in differentiated HL60 cells. *Nature,* **321**, 702.

143. Adhya,S. and Gottesman,M. (1982) Promoter occlusion: transcription through a promoter may inhibit its activity. *Cell,* **29**, 939.

144. Payne,G.S., Courtneidge,S.A., Crittenden,L.B., Fadly,A.M., Bishop,J.M. and Varmus,H.E. (1981) Analysis of avian leukosis virus DNA and RNA in bursal tumors: viral gene expression is not required for maintenance of tumor state. *Cell,* **23**, 311.

145. Neel,B.G., Hayward,W.S., Robinson,H.L., Fang,J. and Astrin,S.M. (1981) Avian leukosis virus-induced tumors have common proviral integration sites and synthesise discrete new RNAs: oncogenesis by promoter insertion. *Cell,* **23**, 323.

146. Cullen,B.R., Lemedico,P.T. and Ju,G. (1984) Transcriptional interference in avian retroviruses—implications for the promoter insertion model of leukaemogenesis. *Nature,* **307**, 241.

147. Proudfoot,N.J. (1986) Transcriptional interference and termination between duplicated

α-globin gene constructs suggests a novel mechanism for gene regulation. *Nature,* **322**, 562.

148. Ju,G. and Cullen,B.R. (1985) The role of avian retroviral LTRs in the regulation of gene expression and viral replication. *Adv. Virus. Res.,* **30**, 179.

149. Emerman,M. and Temin,H.U. (1984) High-frequency deletion in recovered retrovirus vectors containing exogenous DNA with promoters. *J. Virol.,* **50**, 42.

150. Kadesh,T.R. and Berg,P. (1983) In *Enhancers and Eukaryotic Gene Expression.* Gluzman,Y. and Shenk,T. (eds), Cold Spring Harbor Laboratory Press, New York, p. 21.

151. Cook,P.R. and Brazell,I.A. (1975) Supercoils in human DNA. *J. Cell Sci.,* **19**, 261.

152. Ptashne,M. (1986) Gene regulation by proteins acting nearby and at a distance. *Nature,* **322**, 697.

4

RNA splicing

Adrian R.Krainer and Tom Maniatis

1. Introduction

The majority of eukaryotic transcription units are composed of coding sequences, or exons, interrupted by one or more non-coding intervening sequences, or introns. RNA splicing is a series of cleavage and ligation reactions that result in the precise excision of introns from the precursor RNA, and in the joining of adjacent exons, to generate mature RNA. Although several theories have been proposed to explain the function and evolutionary significance of introns, many unanswered questions remain. It is clear, however, that as a consequence of the organization of interrupted genes into exons and introns, RNA splicing is essential for gene expression. Furthermore, in at least some cases, gene expression may be regulated at the level of RNA splicing.

Since the discovery of RNA splicing, the development of cell-free systems capable of splicing well-defined, *in vitro*-generated RNA precursors has allowed many insights into the mechanisms of several different kinds of RNA splicing reactions. Other recent reviews have discussed nuclear pre-mRNA splicing (1,2), alternative splicing (3), group I intron splicing and RNA catalysis (4), small nuclear ribonucleoproteins (5) and heterogeneous ribonucleoprotein particles (6). In this review we discuss these and other topics related to pre-mRNA splicing, including recent advances in the study of small nuclear ribonucleoproteins (snRNPs), splicing complexes and self-splicing introns.

2. Splicing mechanisms

Introns have been found in a wide variety of RNA precursors, including nuclear and organelle pre-mRNA, pre-rRNA and pre-tRNA. Close examination of the structures of these introns, and of the mechanisms for their excision, has shown that several distinct splicing pathways exist. For each of these splicing reactions, a unique mechanism of splice site recognition and of cleavage and ligation has evolved. At present, most known introns can be assigned unambiguously to one of four classes, depending on the intron structure and location. Splicing mechanisms have been partly elucidated for examples of each of the four classes of introns, which include:

(i)   nuclear pre-tRNA introns
(ii)   group I introns
(iii) group II introns
(iv)   nuclear pre-mRNA introns.

   This review is concerned primarily with the mechanism of splicing of nuclear pre-mRNA introns, but the remaining reactions are summarized below for comparative purposes.

2.1 Splicing of nuclear pre-tRNA introns

The mechanism of nuclear pre-tRNA splicing in *Saccharomyces cerevisiae* was the first to be understood in detail (reviewed in ref. 7). Several nuclear-encoded yeast pre-tRNAs contain a single, short intron located at a conserved position, one nucleotide 3′ of the anticodon (see ref. 8 and *Figure 1*). A few animal and plant nuclear pre-tRNAs are also interrupted

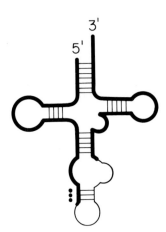

Figure 1. Structural model of a pre-tRNA. The characteristic postulated structure of an interrupted pre-tRNA is shown schematically. The exons are indicated by the thick lines. The dots denote the position of the anticodon, one nucleotide 5′ of the intron. Base pairing of the anticodon and a complementary intron sequence, and the presence of the splice sites in single-stranded loops are characteristic features of *S.cerevisiae* pre-tRNAs, but these features are not required for splicing (9 – 11).

at the same position by a short intron. (Some chloroplast tRNA genes contain much longer introns within their anticodon loops; their sequences suggest that some are group I and others are group II introns; see Sections 2.2 and 2.3). No sequence or strict length conservation is apparent in nuclear pre-tRNA introns, which range in size from 8 to 60 nucleotides (8), and appear to lack essential splicing signals. The intron constitutes a separate structural domain that interrupts the anticodon loop but does not appear to disrupt the overall conserved secondary and tertiary structure of mature tRNA (9,10). Although these conserved features are also involved in other aspects of tRNA biogenesis and function, it appears that recognition of splice sites by the tRNA splicing enzymes is based primarily on common structural features of the exons, and on the conserved position of the intron (9–11, and references therein).

In yeast the pre-tRNA splicing reaction was shown to take place in two steps (12) (*Figure 2*). In the first step, endonucleolytic cleavages at the 5' and 3' splice sites take place, with release of a linear intron. In the second step, the two exons are ligated. The two steps can be uncoupled in the absence of ATP, which is only required for exon ligation. Two factors, one of which is responsible for endonucleolytic cleavage and one for ligation, have been purified (13–15). The endonuclease is an integral nuclear membrane protein that generates 5'-hydroxyl termini and 2',3'-cyclic phosphodiester termini at both cleavage sites (13). The ligase is a peripheral nuclear membrane protein that appears to be capable of several enzymatic reactions necessary for modification of the termini in preparation for the joining of the cleaved exons (14,15) (*Figure 2*); these enzymatic reactions were first described for RNA ligation in wheat germ extracts (16,17). A 3'-cyclic phosphodiesterase activity converts the 2',3'-cyclic phosphodiester terminus into a 2'-phosphomonoester. The exon 2 fragment is phosphorylated at the 5' terminus by a kinase activity that transfers the γ phosphate of an ATP cofactor; this phosphate will form the phosphodiester bond at the spliced junction. The remaining steps are analogous to the mechanism of T4 RNA ligase (18–20), except for the presence of the 2'-phosphomonoester group. The phosphorylated 5' terminus is adenylylated by transfer of an AMP moiety from a ligase-AMP intermediate, forming a 5',5'-diphosphate linkage. The exons are joined via a 2'-phosphomonoester, 3',5'-phosphodiester bond, while AMP is released. Finally, the 2'-phosphomonoester group is removed by an uncharacterized 2'-phosphatase activity, to generate a 2'-hydroxyl group.

A different pre-tRNA splicing mechanism has been proposed for higher eukaryotes, based on the observation that when yeast and *Xenopus laevis* pre-tRNAs are spliced in frog oocyte or HeLa cell extracts, the 3'-terminal phosphate of exon 1 is the one that forms the phosphodiester bond at the spliced junction (21–23) (*Figure 3*). Recently, human and plant nuclear pre-tRNAs were spliced in the HeLa extract, presumably by the same mechanism (24).

A HeLa cell RNA ligase activity has been partially purified and

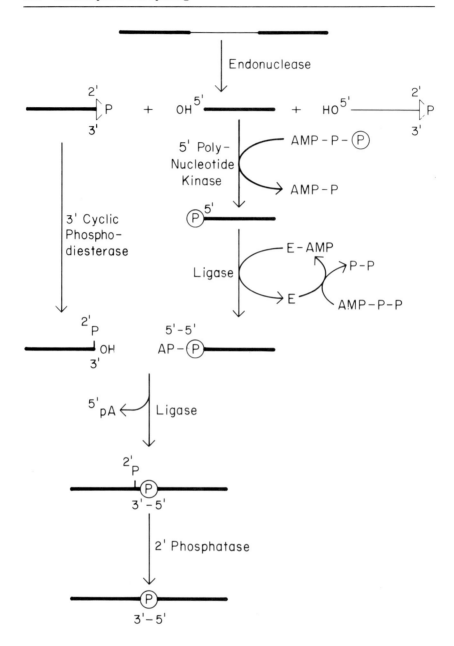

Figure 2. Mechanism of pre-tRNA splicing in *S.cerevisiae*. The pathway of pre-tRNA splicing in *S.cerevisiae* is shown schematically (13–15). The exons are indicated by the thick lines. E-AMP represents an adenylylated enzyme complex. Kinase, phosphodiesterase, and ligase activities are associated with a single 90 kd polypeptide (15). The origin of the spliced junction phosphate, and the nature of the cleaved termini and newly formed bonds are indicated.

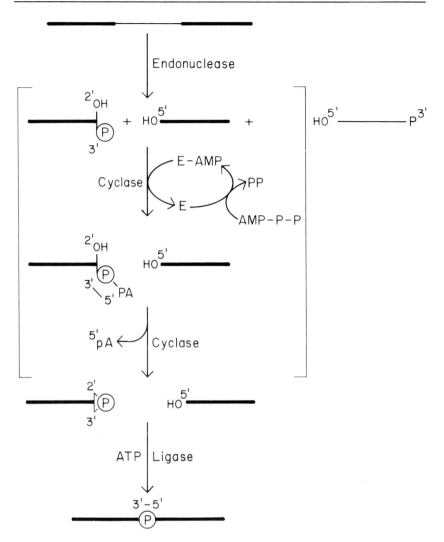

Figure 3. Mechanism of pre-tRNA splicing in HeLa and *Xenopus* extracts. A hypothetical pathway for metazoan pre-tRNA splicing is shown schematically. It is not known whether the endonuclease generates 2′,3′-cyclic phosphodiester termini directly, or whether it generates intermediate 3′ phosphomonoester termini. ATP-dependent cyclase activities that can generate a cyclic terminus from a 3′-phosphomonoester terminus via a 3′ adenylylated intermediate have been described, but their involvement in this pathway has not been demonstrated. The origin of the spliced junction phosphate is indicated. The role of ATP in the ligation step is not understood. The structure of the 3′ terminus of the excised intron has not been well characterized (7,21–27).

characterized. It can join 2′,3′-cyclic phosphodiester and 5′-hydroxyl termini via a conventional 3′,5′-phosphodiester bond derived from the cyclic phosphate, and may be responsible for ligation of tRNA exons

(25,26). Finally, RNA cyclase activities have been found in wheat germ and HeLa extracts. These activities can convert 3'-phosphomonesters to 2',3'-cyclic phosphodiesters via a 3' adenylylated intermediate, and may be responsible for the generation and/or maintenance of the cyclic 3' termini generated during endonucleolytic cleavage (25,27).

2.2 Splicing of group I introns

The mechanism of splicing of transcripts containing group I introns was first elucidated for the macronuclear 26S pre-rRNA of *Tetrahymena* (28,29; reviewed in ref. 4). Splicing of this pre-rRNA *in vitro* was shown to occur in the absence of proteins or an external source of energy; only magnesium and guanosine are necessary. This observation constituted the first demonstration of catalytic activity in an RNA molecule (reviewed in refs 4,30,31). Group I introns have also been found in the nuclear pre-rRNA of *Physarum polycephalum* (32), and in mitochondrial pre-rRNAs and pre-mRNAs of several species of fungi (reviewed in refs 4,33 – 36). Some of the fungal mitochondrial group I introns contain open reading frames encoding proteins known as maturases (37; reviewed in ref. 36). Finally, group I introns have been found in some plant chloroplast tRNA, rRNA and mRNA genes (38; reviewed in ref. 39), and in the pre-mRNA encoding thymidylate synthetase in bacteriophage T4 (40,41). The ubiquitous presence of group I introns in nuclear and organelle genes, in prokaryotes and eukaryotes, and in genes transcribed by many different RNA polymerases, suggests that this type of intron was already present in primordial organisms, prior to the divergence of prokaryotes and eukaryotes, or that these introns are capable of insertion at new locations (42).

A common feature of all group I introns is the presence, in a conserved order along the RNA, of several short sequence elements that exhibit pairwise complementarity (reviewed in refs 4,33 – 36). Four of these elements are conserved in primary sequence, and are known as A, B, 9L and 2 (4), or in an alternative nomenclature, as P, Q, R and S (35). Eight additional stretches of sequence with pairwise complementarity but no primary sequence conservation are also universal. Two of these complementary elements are known as 9R and 9R', or as E' and E. Another of these elements is known as the internal guide sequence (IGS), and is complementary to the 3' border of the 5' exon (35). Several of these conserved elements are known to be important, because they are sites of *cis*-acting splicing mutations (reviewed in refs 4,36). The sequence complementarity between pairs of elements may allow them to interact to form a complex higher order structure, consisting of six helical segments, one of which is at the level of tertiary structure (*Figure 4*). This complex structure appears to be essential for splicing activity. In addition, the interaction between the IGS and the 5' exon is important for 5' splice

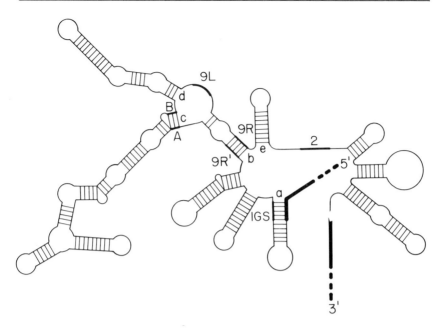

Figure 4. Structural model of a group I intron. The characteristic postulated structure of a group I intron is shown schematically. The 5' and 3' exons are indicated by the thick lines. The conserved elements A, B, 9L, 2, 9R, and 9R' are indicated. The conserved internal guide sequence is denoted by IGS. The five universal secondary structure helices are labeled a – e. A sixth universal helix involves tertiary interactions between elements 9L and 2. Additional secondary structure elements are variable among different group I introns; the structure shown here is based on the structure of the *Tetrahymena* pre-rRNA (reviewed in refs 4,34).

site selection (4,43). The proposed structure has been tested by phylogenetic comparisons, by analyses of compensatory mutations, and by studies with enzymatic and chemical probes. Additional sequence conservation of group I introns includes a U at the 3' border of exon 1, and a G at the 3' border of the intron.

A common splicing mechanism has been demonstrated for several group I introns (reviewed in ref. 4) (*Figure 5*). This splicing reaction requires magnesium and a guanosine cofactor, but it does not require nucleotide hydrolysis as an external source of energy. The energy requirement is obviated by a transesterification mechanism in which the number of phosphodiester bonds formed and broken is the same (4,44). The first transesterification is initiated by covalent joining of the guanosine cofactor. The phosphodiester bond at the 5' splice site is broken and concomitantly, the 5'-phosphate terminus of the intron is esterified with the 3'-hydroxyl group of the guanosine cofactor. This reaction generates a free exon 1 species, and a linear intron still joined to exon 2, and containing a 5'-terminal non-encoded guanosine. These two RNA species were

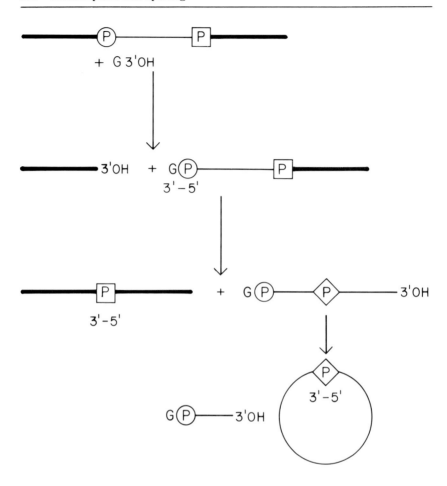

Figure 5. Mechanism of group I intron self-splicing. The typical self-splicing pathway of group I introns is shown schematically. In some cases the excised linear intron has been shown to undergo self-cyclization with release of a small 5'-terminal intron fragment. The required G cofactor is covalently joined to the 5'-terminus of the intron. The fates of relevant phosphates, the structures of the cleaved termini, and the nature of the chemical bonds formed are indicated (reviewed in ref. 4).

recently isolated and shown to be true intermediates (4). In a second transesterification reaction, the phosphodiester bond at the 3' splice site is broken and, concomitantly, the 5'-phosphate terminus of the second exon is esterified with the 3'-hydroxyl group of the first exon. As a result of the two transesterification reactions, the two exons are precisely joined, and the intron, which has acquired a 5'-terminal guanosine, is released in linear form. In a side reaction involving a similar transesterification mechanism, the *Tetrahymena* linear intron can circularize with simultaneous release of a 5'-terminal 15 nucleotide fragment (4,44), supporting the notion that the intron contains all the structural features necessary for the RNA catalyzed reactions (reviewed in ref. 4).

Although the catalytic properties of group I introns are presumably encoded in the conserved sequences and structural features of the intron, some members of this class of introns do not self-splice *in vitro*. For example, the *Neurospora crassa* large rRNA intron, which contains the conserved group I sequence elements and which acquires a 5'-terminal guanosine upon excision *in vivo*, does not appear to self-splice *in vitro* (45). An activity that promotes splicing of this pre-rRNA *in vitro* has been partially purified and appears to be a protein (46). The first intron of the mitochondrial cytochrome *b* gene of *N.crassa* is capable of self-splicing *in vitro* by the standard group I mechanism, but its processing *in vivo* can be blocked by a recessive mutation in a nuclear-encoded gene (45). The same mutation blocks splicing of other group I introns *in vivo*, including the large rRNA intron (46). A maturase-containing group I intron in the yeast mitochondrial *oxi3* gene is capable of self-splicing *in vitro*, but its processing *in vivo* requires mitochondrial protein synthesis (47). Therefore, superimposed on the self-splicing nature of group I introns, some members of this class may require, or be subject to regulation by, nuclear-encoded or intron-encoded proteins. It seems likely that these proteins are involved in determining the overall secondary and tertiary structure of the pre-RNA that is necessary for optimal splicing (48). However, it is formally possible that in some cases, these proteins have enzymatic activities that replace or supplement the intrinsic catalytic activity of group I introns.

An interesting parallel can be made with ribonuclease P from *Escherichia coli* (reviewed in refs 30,31). The M1 RNA subunit of RNase P is the catalytic subunit of the enzyme, and in the absence of the C5 protein subunit, it can correctly process tRNA substrates *in vitro* (49). However, the protein subunit, which has no demonstrable catalytic activity by itself, is required for enzymatic activity *in vivo* (reviewed in ref. 30). Furthermore, the protein subunit is necessary for the *in vitro* processing of 4.5S RNA (30). Current hypotheses for the mode of action of the C5 subunit include a role in electrostatic shielding, and participation in a necessary conformational change of the M1 subunit (reviewed in refs 4,30).

2.3 Splicing of group II introns

Group II introns are found in some pre-mRNAs of fungal and plant mitochondria, and in some chloroplast pre-tRNAs. A few encode maturases (reviewed in refs 34,36,50). This class of introns is characterized by the following consensus 5' and 3' splice sites: /GUGCG and YUAYYNY(N)AY/ (50,51). In addition, six internal putative hairpin structures have been postulated (34,50) (*Figure 6*). The fifth potential

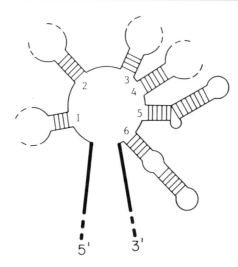

Figure 6.   Structural model of a group II intron. The characteristic postulated structure of a group II intron is shown schematically. The 5' and 3' exons are indicated by the thick lines. Dotted lines indicate RNA segments of variable length and structure that are not conserved among different group II introns (reviewed in ref. 34).

hairpin structure is the most conserved in sequence and structure, and is located within 100 nucleotides of the 3' end of the intron. The conserved 3' splice site, except for the last two or three bases, forms part of the sixth hairpin. Finally, the length and sequence of the unpaired segments connecting adjacent hairpins are also conserved. Mutations in several of the proposed hairpins have effects on splicing *in vivo* (52 and references therein) and *in vitro* (52).

Recent work on the mechanism of splicing of the seventh intron of the yeast mitochondrial *oxi3* gene showed that this group II intron is excised autocatalytically, but in a manner that differs substantially from the self-splicing of group I introns (53,54). The optimized *in vitro* reaction requires magnesium and spermidine, but no nucleoside cofactor. Surprisingly, the reaction appears to take place via a lariat intermediate. This type of branched RNA molecule was previously identified as a distinctive intermediate in the splicing of nuclear pre-mRNA, a reaction that requires cellular proteins and RNAs (see Section 2.4, and *Figure 8*).

Subsequently, it was shown that another group II intron, the first intron of the yeast mitochondrial apocytochrome *b* gene, also self-splices under similar conditions (52). In both cases, the site of branch formation appears to be eight nucleotides or so upstream of the 3' end of the intron, within the conserved 3' splice site sequence (52,54). The branched intermediates have been identified by several criteria, but the branch nucleotide has not been directly identified, and the nature of the chemical bonds remains to be precisely determined. However, these intermediates are almost certainly analogous to those generated during splicing of nuclear pre-mRNA. Since no external source of energy is required, this mechanism must involve transesterification reactions.

The catalytic activity and specificity of the reaction must be at least partly determined by the conserved primary and secondary structures of the group II introns. Thus, mutations in these structures affect self-splicing (52). However, as in the case of some group I introns, splicing of some group II introns *in vivo* requires and/or is modulated by nuclear-encoded factors (55), and sometimes by maturases (56).

2.4 Splicing of nuclear pre-mRNA introns

With few exceptions, nuclear pre-mRNAs are interrupted by introns, ranging in number from one to fifty. The length of known natural pre-mRNA introns varies from 31 nucleotides (57) to approximately 60 kb (58,59). Three conserved sequence elements have been identified in the nuclear pre-mRNA introns of higher eukaryotes, at the 5' and 3' splice sites (60,61), and at a site within the intron near the 3' splice site, known as the branchpoint sequence (62–64) (*Figure 7*). Recently derived consensus sequences from plant introns show strong similarities to the animal consensus elements (65) (*Figure 7*). In *S.cerevisiae*, a few pre-mRNAs contain one and, in one known case (66), two introns. In contrast to the animal and plant sequence elements, which can deviate from the consensus sequences at several positions, the analogous sequence elements in *S.cerevisiae* show very little deviation from their respective consensus sequences (67–69) (*Figure 7*). In the fission yeast *Schizosaccharomyces pombe*, introns appear to be more common than in *S.cerevisiae* and to have more flexible sequence requirements for splicing (reviewed in ref. 70).

Analyses of the *in vivo* and *in vitro* effects of naturally occurring and *in vitro*-generated mutations demonstrated the importance of the conserved 5', 3', and branchpoint sequences for accurate and efficient splicing in animal cells and in *S.cerevisiae* (reviewed in refs 1,2,71,72; see Section 4.2). The relative contributions of these elements in different systems are addressed by studies of the processing of animal, plant or fungal pre-mRNAs in heterologous systems *in vivo* and *in vitro* (73–80). The three conserved elements in the pre-mRNAs of animal cells and their viruses appear to be insufficient to account for the high degree of specificity of the reaction. The mechanisms responsible for the specificity of splice site selection are poorly understood (see Section 4).

The nuclear pre-mRNA splicing pathway, as presently understood from studies in which pre-mRNAs containing a single intron are spliced in HeLa or *S.cerevisiae* cell-free systems, occurs as follows. The pre-mRNA associates with multiple components to form a large complex known as a spliceosome (81–83; see Section 6.4). Formation of the complex requires

ATP, functional splice sites, snRNPs, heterogeneous nuclear ribonucleo-proteins (hnRNPs) and other protein factors. Some of these factors are stable components both of the spliceosome and of a number of smaller complexes that may represent intermediates in spliceosome formation. Additional extrinsic factors appear to interact with the complexes in a transient manner, and some may be necessary for proper spliceosome assembly. In mammalian extracts, assembly by these complexes takes place during a time lag that precedes the cleavage and ligation reactions (84 – 86). In yeast extracts, on the other hand, no significant time lag has been observed (87).

Following the assembly of the 60S splicing complex, the pre-mRNA is cleaved at the 5' splice site (86,88), generating an exon 1 RNA fragment that contains a 3'-hydroxyl group (62,89) (*Figure 8*). The phosphate group at the 5' terminus of the intron is esterified with the 2'-hydroxyl group

Vertebrate 5' splice site consensus:

$$\genfrac{}{}{0pt}{}{C_{38}}{A_{39}}\ A_{62}\ G_{77}\ \Big|\ G_{100}\ U_{100}\ A_{60}\ A_{74}\ G_{84}\ U_{50}$$

Vertebrate 3' splice site consensus:

$$Y_{77}\ Y_{78}\ Y_{81}\ Y_{83}\ Y_{89}\ Y_{85}\ Y_{82}\ Y_{81}\ Y_{86}\ Y_{91}\ Y_{87}\ N\quad Y_{97}\ A_{100}\ G_{100}\ \Big|\ G_{55}$$

Mapped vertebrate and viral natural branchpoints:

$$\underset{16}{Y_{13}}\ N\quad \underset{16}{Y_{16}}\ \underset{16}{U_{14}}\ \underset{16}{R_{13}}\ \underset{16}{\overset{\bullet}{A}_{16}}\ \underset{16}{Y_{15}}\ \text{----}\quad A\ G\ \Big|$$

$\longleftarrow$ 18 – 37 nt $\longrightarrow$

Plant 5' splice site consensus:

$$\genfrac{}{}{0pt}{}{C_{33}}{A_{33}}\ A_{55}\ G_{72}\ \Big|\ G_{100}\ U_{100}\ A_{70}\ A_{55}\ G_{65}\ U_{49}$$

Plant 3' splice site consensus:

$$Y_{70}\ Y_{69}\ Y_{71}\ Y_{56}\ Y_{66}\ Y_{60}\ Y_{61}\ Y_{53}\ Y_{56}\ Y_{57}\ Y_{78}\ R_{70}\ Y_{94}\ A_{100}\ G_{100}\ \Big|\ G_{60}$$

S. cerevisiae 5' splice site consensus:

$$\Big|\ \underset{23}{G_{23}}\ \underset{23}{U_{23}}\ \underset{23}{A_{23}}\ \underset{23}{U_{20}}\ \underset{23}{G_{23}}\ \underset{23}{U_{22}}$$

S. cerevisiae 3' splice site consensus:

$$\underset{23}{Y_{23}}\ \underset{23}{A_{23}}\ \underset{23}{G_{23}}\ \Big|$$

S. cerevisiae branchpoints:

$$\underset{23}{U_{23}}\ \underset{23}{A_{23}}\ \underset{23}{C_{23}}\ \underset{23}{U_{23}}\ \underset{23}{A_{23}}\ \underset{23}{\overset{\bullet}{A}_{23}}\ \underset{23}{C_{23}}\ \underset{23}{A_{19}}\ \text{----}\quad A\ G\ \Big|$$

$\longleftarrow$ 10 – 139 nt $\longrightarrow$

of the conserved adenosine residue within the branchpoint sequence. This forms a branched circular RNA, termed a lariat (62,89). This first cleavage – ligation reaction appears to occur in a concerted manner, and hence, by analogy to the transesterification reactions characteristic of group I intron splicing, it may not require external energy (42). Although pre-mRNA splicing does require ATP (84 – 86) the exact role of this nucleotide in the overall reaction has not been elucidated (see Section 5.2).

Two intermediates are generated by this cleavage – ligation reaction: an exon 1 fragment and an intron – exon 2 fragment with the intron in the form of a lariat (*Figure 8*). These two RNAs are held together non-covalently, as part of the spliceosome (82,83). In the next step of the reaction, the intron – exon 2 intermediate is cleaved at the 3′ splice site, generating a free intron lariat containing a 3′-hydroxyl group (62,89). This cleavage reaction appears to occur in concert with the ligation of the two exons, which occurs by esterification of the 5′-terminal phosphate group of the exon 2 fragment with the 3′-terminal hydroxyl group of the exon 1 fragment (62,89). These two cleavage – ligation reactions will be referred to as the 5′ and 3′ splice site reactions, respectively, although the former reaction involves the branchpoint sequence, which is located near the 3′ splice site (*Figure 7*).

In both yeast and mammalian *in vitro* systems, the two cleavage – ligation steps can be uncoupled from each other under defined biochemical conditions, or by selective inactivation or removal of at least two activities that are only necessary for the second reaction (90 – 93). Abortive intermediates that are blocked before the 3′ splice site reaction

Figure 7. Conserved sequence elements in pre-mRNA introns. Vertebrate 5′ and 3′ splice site consensus sequences (61) are derived from a recent compilation of approximately 400 introns (1). The subscripts denote the frequency of occurrence of the consensus bases at each position, expressed as a percentage. A C residue at position 2 of the 5′ splice site has been described in three natural introns (293 – 295). A C residue at the last position of the 3′ splice site has been found in one natural *Drosophila* intron (296). The metazoan branchpoint consensus (62 – 64) is derived from 16 wild-type vertebrate and viral branchpoints mapped after processing in a HeLa extract or *in vivo* (62,63,94,136,159,160,203,297). The subscripts denote the frequency of occurrence of the consensus bases at each position. The branchpoint A residue is indicated by a dot, and the range of distances to the 3′ splice site is indicated. The plant 5′ and 3′ splice site consensus sequences (65,78) are derived from a recent compilation of 177 introns (65). The subscripts denote the frequency of occurrence of the consensus bases at each position, expressed as a percentage. Although a plant putative branchpoint consensus has been derived (65), actual plant branchpoints have not been mapped after processing in homologous systems. The *S.cerevisiae* consensus sequences (67 – 69) are derived from 23 introns, including 17 ribosomal protein introns (16 sequences compiled in ref. 67; others from 68,298 – 301). The subscripts denote the frequency of occurrence of the consensus bases at each position. The branchpoint A residue is indicated by a dot, and the range of distances to the 3′ splice site is indicated; the range is 22 – 58 nt if the MatA1 and Tub3 introns are excluded. The 5′ and 3′ splice sites are indicated by vertical lines.

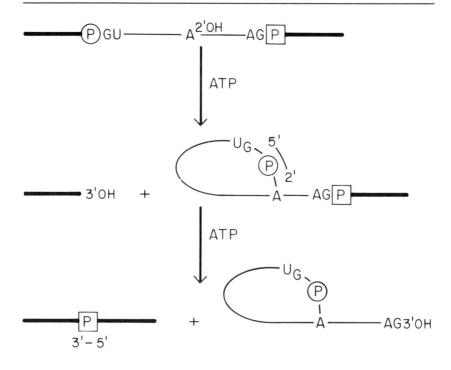

Figure 8.   Mechanism of nuclear pre-mRNA splicing. The pathway of nuclear pre-mRNA splicing, as demonstrated in metazoans and in *S.cerevisiae*, is shown schematically. The roles of ATP in both steps of the reaction have not been elucidated. The fates of relevant phosphates, the structures of the cleaved termini, and the nature of the chemical bonds generated are indicated. This reaction requires multiple factors, including proteins and snRNPs. Similar intermediates are generated during the self-splicing of group II introns, but this reaction requires only magnesium and spermidine.

can also be obtained as a result of point mutations in the AG dinucleotide at the 3′ splice site (94,95), and in some but not all cases, by point mutations in the GU dinucleotide at the 5′ splice site (96 – 99), or in the branchpoint adenosine (97,98,100,101).

Multiple factors are necessary for nuclear pre-mRNA splicing in mammalian systems (90,91,102,103; see Sections 6.1 – 6.3). Several of these are snRNP particles, which consist of one or two small nuclear RNAs and multiple polypeptides (reviewed in refs 5,104; see Section 6.1). All major mammalian nucleoplasmic snRNPs are thought to be involved in pre-mRNA splicing. The evidence is most extensive in the case of U1 snRNP, whose ability to interact with 5′ splice sites by RNA – RNA base pairing is now known to play an important role in splice site selection (105). U2 snRNP (90,106) and the U4/U6 snRNP complex (107,108) are also required for splicing. U2 snRNP appears to interact specifically with the pre-mRNA branchpoint region under splicing conditions (106). Another snRNP that may be U5 can interact specifically with 3' splice sites under splicing conditions (109). In addition, the snRNAs are stably associated with the spliceosome (82,110–113). Other factors that appear to be proteins are also necessary for pre-mRNA splicing *in vitro* (90,91,102,103).

Pre-mRNA splicing in yeast also requires multiple factors (92,114). Some of these have been partially purified, but their identity is unknown, although two fractions appear to contain essential snRNPs. Several of the yeast snRNPs are stably associated with the splicing complexes (115). snRNPs were previously known to exist in yeast, but their functional relationship to the mammalian snRNPs was uncertain (116). A large RNA known as LSR1 (117) or snR20 (118), which bears partial homology to metazoan U2 snRNA (117), and is part of the spliceosome (115), participates in the recognition of the branchpoint sequence, since mutations at this site can be rescued by compensatory changes in the cloned snR20 gene (119). In addition, several more required components have been identified recently by biochemical complementation between extracts derived from different splicing-defective temperature-sensitive mutants in the *RNA* loci. The products of the genes *RNA2*, 3, 4, 5, 7, 8 and 11 are necessary for splicing (114), and all appear to be necessary for spliceosome formation (93). The *rna2* gene product can form an intermediate splicing complex that is blocked prior to the 5' splice site reaction (93). The combined genetic and biochemical approaches should greatly facilitate the identification, purification and characterization of the components of the yeast pre-mRNA splicing machinery.

2.5 Comparison of group I, group II and nuclear pre-mRNA splicing mechanisms

The mechanistic parallels among the different types of splicing reactions, excluding tRNA splicing, were pointed out in recent reviews (1,42). The similarity is most obvious in the case of group I and group II intron splicing mechanisms, since both reactions involve autocatalytic transesterifications (53). The first transesterification in the case of group I splicing is initiated by a nucleophilic attack by the 3'-hydroxyl group of the guanosine cofactor, which is esterified with the 5'-terminal phosphate of the intron, resulting in simultaneous cleavage at the 5' splice site. In the case of group II introns, no nucleoside cofactor is necessary; instead the nucleophile appears to be the free 2'-hydroxyl group of a nucleotide within the RNA. Since all the nucleotides in the RNA contain available 2'-hydroxyl groups, how a single nucleotide within the intron is chosen for this role is not known. This choice is presumably determined by the sequence around the branchpoint and/or the overall higher order structure of the RNA. Group I introns contain a binding site for the external guanosine cofactor

(120). By analogy, folding of the group II introns should position the branchpoint nucleotide within the intron active site.

Following the first transesterification reaction, group I introns are linear and contain a 5′-terminal non-encoded guanosine, whereas group II introns are in lariat form. In both cases the introns remain joined to the 3′ exon. In the second transesterification step both mechanisms involve nucleophilic attack by the 3′-hydroxyl group of the 5′ exon, which becomes esterified with the 5′ phosphate at the 5′ terminus of the 3′ exon. This reaction results in cleavage at the 3′ splice site, releasing the intron, and in simultaneous ligation of the exons.

In both mechanisms, a released 5′ exon molecule containing a 3′-hydroxyl group is generated, and the intron remains covalently joined to exon 2. An important aspect of the mechanisms, which can be addressed experimentally, is the identification of the binding site for the non-covalently attached exon 1 within the intron–exon 2 molecules. A sequence within group I introns, known as the internal guide sequence (35), appears to base pair in a functionally significant manner with a complementary sequence in the 5′ exon, immediately upstream of the 5′ splice site (reviewed in ref. 4). The internal guide sequence is thought to be part of the active site for both transesterification steps (reviewed in ref. 4). The base-pairing interaction is important both for recognition of the 5′ splice site prior to the first transesterification, and for binding and alignment of the 5′ exon leading to 3′ splice site cleavage and exon ligation (43).

A 5′ exon binding site within a group II intron has been suggested, but not localized, by *trans* self-splicing experiments with separate 5′ exon and intron–3′ exon linear RNAs (121). Interestingly, in these *trans*-splicing experiments 5′ exon binding, as well as intron-catalyzed 5′ splice site cleavage activity, exon ligation and intron excision, could take place with the intron in either linear or lariat form (121). This observation suggests that the catalytic structure of this group II intron does not change significantly when its 5′ terminus is covalently joined to a nucleotide very close to its 3′ terminus.

Recently, the self-splicing *oxi3* group II intron was shown to require 5′ exon sequences (122). Secondary structure models that involve base-pairing between the 3′ terminus of the 5′ exon and an internal guide sequence were proposed, but these putative structures did not appear to be conserved in other group II introns (122).

A 5′ exon binding site probably also exists within the splicing complexes formed with nuclear pre-mRNAs, since the two RNAs generated by cleavage at the 5′ splice site and lariat formation are part of the same ribonucleoprotein complex (81–83). The 5′ terminus of U1 snRNA interacts with the 5′ splice site by base-pairing (105,123,124), which is reminiscent of the above-mentioned base-pairing interaction between the internal guide sequence of group I introns and the exon sequences just upstream of the 5′ splice site. In fact, both kinds of interaction play a role in 5′ splice site selection (43,105). The binding of the cleaved 5′ exon during nuclear pre-mRNA splicing may be accomplished by adjacent sequences within U1 snRNA, or by RNA–protein interactions with U1 snRNP or other components.

These comparisons serve to point out many apparent mechanistic similarities in the splicing of group I and group II introns, even though precise mechanistic details are still unknown, especially in the case of group II introns. Despite these similarities, no sequence or structural homology between group I and group II introns has been found (34). Determination of the secondary and tertiary structure of group I introns, in combination with mutational and biochemical analyses, are helping to identify the active sites and to understand the nature of the RNA–RNA interactions that take place at these sites (reviewed in ref. 4). When similar analyses are carried out for group II introns, it may be possible to make meaningful comparisons of the structures and catalytic activities of the two classes of introns.

Some group I and group II mitochondrial introns contain open reading frames that encode polypeptides known as maturases (reviewed in ref. 36). These protein products are known to be necessary for splicing from genetic studies, which show that splicing-defective mutations in the intron can be complemented in *trans* (37,56,125–128). At least one maturase-containing group I intron is capable of self-splicing *in vitro*, even though its processing *in vivo* requires mitochondrial protein synthesis (47). Other group I and group II introns that encode maturases may or may not self-splice *in vitro*. Perhaps some maturase-encoding introns have evolved in such a way that some of the structural and/or catalytic functions of the intron itself have been transferred to the maturase protein. For example, the maturase may be necessary for proper folding of the RNA. The existence of maturases could provide an evolutionary advantage in some cases, because it provides new opportunities for regulation. For example, it has been proposed that maturases encoded by open reading frames that span an exon–intron border are subject to feedback regulation, since only unspliced RNA can encode functional maturases (37).

Even group I and group II introns that do not encode maturases may be subject to regulation by proteins *in vivo*. For example, a splicing mutation in the first intron of yeast mitochondrial apocytochrome *b*, which self-splices *in vitro* by the group II pathway, can be suppressed by an allele-specific suppressor mutation in a nuclear-encoded gene (52,55). This observation strongly suggests that the nuclear-encoded product is a splicing factor that interacts with intron sequences, and appears to be necessary for a normal rate of splicing *in vivo*. Similarly, the first intron of the mitochondrial cytochrome *b* gene of *N.crassa* self-splices *in vitro* by the standard group I mechanism, but its processing *in vivo* can be blocked by a recessive mutation in a nuclear-encoded gene (45). The

N.crassa group I intron in the mitochondrial large pre-rRNA requires a nuclear-encoded protein for splicing both *in vivo* and *in vitro* (46).

Despite the apparent similarities between the lariat intermediates generated during nuclear pre-mRNA splicing and group II splicing, the former reaction requires multiple factors and an external source of energy. It remains possible that under different conditions, nuclear pre-mRNAs could splice autocatalytically. However, the extremely large number of different nuclear pre-mRNA introns, and the lack of strong internal sequence conservation, suggest that external factors, rather than the introns, may be responsible for proper folding of the pre-mRNA leading to correct selection of splice sites. In addition, these factors may actually catalyze the cleavage – ligation reactions. An important distinction can be made between a structural role, in which a factor accelerates the reaction rate by promoting proper folding of the substrate, and a catalytic role, in which the factor provides the active site where the actual cleavage and ligation reactions take place. In addition to cleavage and ligation activities, factors may have other catalytic roles, such as hydrolyzing ATP, which is essential for splicing *in vitro*.

The involvement of snRNAs in nuclear pre-mRNA splicing is particularly intriguing, because these RNAs may contain the active sites for the cleavage – ligation reactions. If so, in both types of reaction the catalysis would involve primarily RNA – RNA interactions, whether in *cis* or in *trans*. A precedent of an RNA cleavage reaction that is catalyzed by a separate RNA is the processing of *E.coli* tRNAs by the M1 RNA subunit of RNase P (49). Alternatively, the snRNAs may be necessary only for splice site recognition, which may be accomplished by base-pairing (105,123,124); the active sites may be present in protein factors, or in the pre-mRNA itself. Some of the factors, such as hnRNP proteins, may be required for proper folding of the pre-mRNA. Utilization of such factors would relieve some of the sequence requirements necessary for proper RNA folding in the absence of proteins. However, some sequence or structural conservation is necessary to allow proper recognition by these factors.

3. The pathway of mRNA synthesis

The synthesis and translation of mRNA in prokaryotes are processes that take place in a single compartment and can therefore follow closely in time, such that nascent transcripts can be simultaneously translated by several ribosomes. In eukaryotes, the generation of translatable mRNA involves several discrete steps, including synthesis of the primary transcript, capping, methylation of selected adenosine residues, hnRNP assembly, polyadenylation, splicing and transport to the cytoplasm (reviewed in ref. 129). Some of these steps are discussed below in terms of their temporal and mechanistic relations with pre-mRNA splicing.

3.1 Capping

All eukaryotic nuclear pre-mRNAs are transcribed by RNA polymerase II, which is also responsible for the synthesis of many small nuclear RNAs. These RNAs are modified post-transcriptionally by guanylyl transfer and methylation to form the 5'-terminal cap structure characteristic of RNA polymerase II transcripts (reviewed in refs 130,131). Capping enzyme, or guanylyl transferase, has been purified and studied from several mammalian, plant, fungal and viral sources. An associated triphosphatase activity removes the γ phosphate at the 5' terminus of the RNA to generate an intermediate 5'-diphosphate terminus. The β phosphate at the RNA 5' end is joined to the α 5'-phosphate of GTP, and pyrophosphate is simultaneously released. The result is an unusual 5'–5' triphosphate linkage between the guanosine and the first transcribed base. An S-adenosyl methionine cofactor donates a methyl group for modification of the guanosine at its N7 position by an associated methyltransferase activity. Other methyltransferase activities are responsible for variable internal methylations of the cap, for example 2' O-methylations of one or more riboses.

Capping is known to occur at the site of transcription initiation (132,133), and probably occurs during transcription of the first 20–80 nucleotides (133). The tight coupling between transcription initiation and capping is not obligatory, since purified capping enzyme can modify purified RNAs (reviewed in ref. 130), and under some conditions uncapped RNA polymerase II transcripts can be obtained (132,133).

The role of the cap in translation has been extensively documented (131), but this structure could also play a role in other RNA processing reactions, e.g. splicing and/or polyadenylation. In addition, the cap structure appears to be necessary for the nuclear and cytoplasmic stability of RNA polymerase II transcripts (134,135).

The role of the pre-mRNA cap in splicing has been examined with in vitro systems. Uncapped pre-mRNA can be spliced accurately and efficiently, in both mammalian (86) and S.cerevisiae (87) systems. In HeLa nuclear extracts, a fraction of the uncapped RNA is partially degraded from the 5' end, suggesting the action of endogenous 5' to 3' exonuclease activities (86,136,137). Such activities have been purified from mammalian nuclei (138). Capped RNA is protected from this nuclease or nucleases either because the cap itself confers resistance, or because nuclear cap-binding proteins (139) render the RNA 5' end inaccessible to nucleases. Although a fraction of the uncapped RNA is subject to 5' to 3' degradation, spliced RNA is still efficiently generated (86). It has been shown that these spliced RNAs do not contain a 5' cap, which could have been added by an endogenous guanylyl transferase activity (140).

In contrast to the experiments described above with nuclear extracts, when uncapped pre-mRNAs are spliced in a whole cell extract, no spliced mRNA is detected (141). However, due to the low efficiency of splicing in whole cell extracts (85), a relatively small decrease in splicing efficiency in the absence of a cap may have lowered the signal to below the detection threshold.

The functional involvement of cytoplasmic cap-binding proteins in translation was first suggested by experiments in which cap analogs were shown to inhibit translation of some mRNAs, presumably by competing for binding of these polypeptides. Nuclear cap-binding proteins have been identified by affinity-labeling (139), and they may be involved in pre-mRNA processing, stability, or transport.

High levels of cap analogs fail to inhibit splicing in nuclear extracts (140,142). In contrast, inhibition can be obtained in a less efficient whole-cell extract (141). Pre-incubation of the nuclear extract also renders it sensitive to inhibition by these analogs (142). Unexpectedly, the cap analogs do not have to be present during the pre-incubation to allow subsequent inhibition. Inhibition is at the level of splicing – complex formation (143).

The above results suggest that some kind of change takes place during the pre-incubation, which enables an essential, unidentified component to bind cap analogs. A reasonable assumption is that this putative component is originally bound to endogenous RNAs containing capped 5' termini. The extracts may contain endogenous pre-mRNA, and factors bound to it may need to be recycled in order to splice exogenous pre-mRNA (90). Alternatively, the cap-binding factor is bound to the trimethyl caps at the 5' termini of snRNAs. Upon pre-incubation, which is known to increase the accessibility of snRNA 5' termini (90,106), this putative factor may be able to bind cap analogs, even those that lack a 2,2-dimethyl group, and such binding would result in splicing inhibition. How such a factor – snRNP cap interaction could be involved in splicing is unclear. Therefore, the cap analog inhibition experiments may reflect the involvement in splicing of either snRNA trimethyl caps, or of pre-mRNA monomethyl caps.

In summary, the pre-mRNA cap structure is not strictly required for splicing *in vitro*. Protection of the RNA 5' terminus from 5' to 3' exonuclease degradation may be its sole physiological role in the nucleus. Although the pre-mRNA cap is not required for efficient splicing *in vitro*, it remains possible that it is involved in the splicing mechanism *in vivo*. The demonstration of a possible involvement of the nuclear cap-binding proteins in splicing or in other aspects of mRNA biogenesis will have to await further experiments.

3.2 Polyadenylation

The manner in which RNA polymerase II terminates transcription is poorly understood, but in some cases there appears to be gradual termination over several hundred base pairs (reviewed in ref. 144; see

Chapter 3). The mature 3' ends of most pre-mRNAs are generated by endonucleolytic cleavage a short distance downstream of a conserved AAUAAA sequence element (reviewed in ref. 144; Chapter 3). A stretch of up to 200 A residues is then polymerized onto the cleaved 3' end. Cleavage at the 3' end and polyadenylation can be uncoupled *in vitro* (145). The downstream cleavage product can be recovered, and contains a 5'-phosphate terminus. The 3' cleavage and polyadenylation can be uncoupled from transcription, since purified RNAs can be accurately processed (145). It has been reported that polyadenylation requires a capped 5' terminus, but the experiments are complicated by the instability of the uncapped RNA (146).

Pulse-chase experiments show that in at least some cases, poly-adenylation precedes splicing *in vivo* (reviewed in ref. 129). When *in vivo* poly(A) addition is inhibited with cordycepin, splicing can still take place (147), indicating that polyadenylation is not required for splicing. Splicing of long pre-mRNAs can take place on nascent transcripts, before RNA polymerase II has transcribed past the polyadenylation site, indicating that 3' end cleavage is not necessary for splicing *in vivo* (148). Splicing can take place *in vitro* in the absence of 3' cleavage or polyadenylation (86). Conversely, the 3' end formation reaction does not require prior intron removal *in vivo* (149,150) or *in vitro* (151).

The enzyme or enzymes responsible for 3' cleavage and polyadenylation activities have not been identified. *In vitro* the RNA substrates for 3' cleavage and polyadenylation are assembled into ribonucleoprotein complexes (1), whose relation to splicing complexes is uncertain. The proposed involvement of U4/U6 snRNP (152) or of other snRNPs in this reaction has not been strongly substantiated experimentally (107). Inhibition of the *in vitro* reaction in the presence of anti-snRNP and anti-U1 antisera, but not non-immune antiserum, has been reported (151). However, it is difficult to rule out the possibility that the inhibition is caused by variable levels of serum proteases. Specific binding to some but not all AAUAAA polyadenylation signals by one or more snRNPs containing trimethyl cap and Sm determinants has been reported (153). However, although micrococcal nuclease treatment of 3' processing extracts abolishes the reaction (153), full activity can be restored by the addition of non-specific RNA (154), suggesting that no endogenous RNAs are directly involved in this reaction.

3.3 HnRNP assembly

The total population of pre-mRNAs in the nucleus, known as heterogeneous nuclear RNA or hnRNA, is found in association with a defined set of abundant proteins, forming ribonucleoprotein complexes known as hnRNP (reviewed in refs 6,155). HnRNP complexes can be visualized by electron microscopy, and display a characteristic structure consisting of beads, approximately 30 nm in diameter, on a string composed of RNA. Intact hnRNP complexes sediment at 50–300S in sucrose gradients. However, the intrabead linker RNA is sensitive to ribonucleases, and hence hnRNP is usually isolated as a monomer or core with a sedimentation value of 30–40S. The core is made up of 80% protein and 20% RNA, and contains approximately 500 nucleotides of RNA and several proteins. The most abundant of these proteins are found in a fixed stoichiometry, and consist of a sextet of polypeptides whose names and apparent molecular weights are as follows: A1 (34 kd), A2 (36 kd), B1 (37 kd), B2 (38 kd), C1 (41 kd) and C2 (43 kd) (6). Immunoprecipitation with monoclonal antibodies and *in vivo* UV-crosslinking experiments have confirmed the association of these proteins with RNA, and also identified other components migrating as doublets of approximately 68 kd and 120 kd (reviewed in ref. 6).

The assembly of hnRNP occurs on nascent transcripts, and hence hnRNP may be the actual substrate for splicing and 3' end formation *in vivo* (reviewed in refs 6,155). HnRNP structure can be reconstituted *in vitro* from purified proteins and RNA (reviewed in ref. 6), but there could be important differences in the precise structures assembled with full length, as opposed to nascent, transcripts. Purified RNAs can serve as splicing substrates *in vivo* (135) and *in vitro* (84), and hence if hnRNP structure is important for splicing, assembly on a nascent transcript is not a prerequisite. A different set of proteins is found associated with cytoplasmic mRNA as opposed to nuclear hnRNA, and therefore hnRNP organization could be involved in transport of mRNA to the cytoplasm, instead of, or in addition to, the proposed involvement in RNA processing (reviewed in ref. 6). Another possibility is that the organization of hnRNA into hnRNP serves primarily a packaging role.

Recent experiments involving inhibition of splicing *in vitro* by incubation (156,157) or depletion (156) of extracts with anti-hnRNP antibodies are consistent with the involvement of hnRNP proteins, particularly the C proteins, in splicing (see Section 6.3). The mode of action of these proteins could be to fold (or unfold) the RNA into a defined RNP structure that is the actual substrate for the processing enzymes. It is even possible that some of these proteins actually have enzymatic activities.

3.4 Excision of multiple introns

Most, but not all, pre-mRNAs contain introns. Exceptions include certain heat shock, histone and interferon transcripts. Splicing can be uncoupled from transcription (135) and does not require a 5' cap (86), 3' end cleavage (86) or poly(A) addition (147). However, capping and usually polyadenylation precede splicing *in vivo* (149). Whereas capping is essentially coupled to transcription initiation (133), and polyadenylation is detected within 90–120 sec after RNA synthesis (149), it appears that splicing of at least some introns takes several minutes (148,158–160). Splicing of all nuclear pre-mRNA introns so far studied takes place by the same pathway, involving lariat intermediates and excision of intact introns. No evidence of stepwise excision of an intron to generate bonafide functional mRNA, as opposed to abortive intermediates, has been reported.

The excision of multiple introns from a single pre-mRNA may proceed by an ordered pathway. Different introns are probably excised with somewhat different kinetics (148,159,161), but it remains to be determined whether at least in some cases there is an obligatory order of excision. Studies in which a subset of the introns are precisely removed, by replacement of a genomic DNA fragment with the corresponding cDNA segment, in order to determine possible effects on the excision of the remaining introns, are difficult to interpret unambiguously. The reason is that the intron sequences may in some cases contain elements that regulate transcription initiation, or influence the stability of the transcript in the nucleus. Furthermore, such transcripts may be assembled into hnRNP structures that are very different from those formed with natural pre-mRNA, with possible consequences for stability, processing, or transport.

3.5 Nuclear-cytoplasmic transport

Mature mRNAs are transported across the nuclear membrane to the cytoplasm. Pre-mRNAs or partially processed mRNAs are retained in the nucleus (reviewed in ref. 129). The mechanisms responsible for transport, and for the discrimination between processed and unprocessed RNAs, are not known. None of the processing reactions discussed above appear to be obligatorily coupled to transport, since they can take place in cell-free extracts that probably lack nuclear membrane fragments. However, coupling to transport could be accomplished if the processing enzymes were tightly associated with the nuclear membrane. For example, the yeast tRNA splicing endonuclease and ligase enzymes are associated with the nuclear membrane (13). No evidence for such an association exists for mRNA processing activities. However, in most cases, localization studies must await the identification of many of the relevant factors. The availability of antibodies that recognize a variety of snRNP determinants has made it possible to study the localization of snRNPs by indirect immunofluorescence microscopy (reviewed in refs 5,104). These particles are found in the nucleoplasm, in no obvious association with the nuclear membrane. A similar location has been found for the hnRNP proteins (reviewed in ref. 6). Pre-mRNA and splicing complexes appear to be associated with the nuclear matrix (see ref. 162 and references therein), and this association may be responsible for the retention of precursors and intermediates of splicing in the nucleus.

A formal possibility is that pre-mRNA contains nuclear retention signals that are removed in the processing pathway, allowing the transport of the mature mRNA. Such signals could be present for example in introns.

This would be consistent with the observed nuclear accumulation of pre-mRNA containing splicing mutations (163). Moreover, some natural pre-mRNAs lack introns, and are nevertheless efficiently transported. In some cases, cDNAs cloned in expression vectors and transfected into mammalian cells do not give rise to cytoplasmic mRNA unless a homologous or heterologous intron is built into the transcription unit (164 – 166). However, this is not a general phenomenon (167,168), and it could be due to RNA instability, rather than to a block in transport. Therefore, it appears that at least in some cases, pre-mRNA splicing and transport to the cytoplasm are not obligatorily coupled.

Studies of ribonucleoprotein structure indicate that nuclear pre-mRNA and cytoplasmic mRNA are associated with different sets of proteins (reviewed in ref. 6). Hence, maturation and transport are accompanied by a change in ribonucleoprotein composition and organization. However, the exact order of these events has not been determined. It may be that the RNP structure is what determines nuclear or cytoplasmic localization. Although some RNAs transcribed from transfected cDNA templates may still assemble into transportable mRNPs, in other cases it may be that the appropriate mRNP structure can only be assembled in the course of maturation.

4. Splice site selection

4.1 Specificity

The excision of intervening sequences and ligation of exons are processes that have to be carried out with great accuracy. This is particularly true in the case of protein-coding mRNAs, because the translational reading frame has to be preserved across each spliced junction. Therefore, the signals that are recognized by the splicing machinery have to be sufficiently unambiguous to allow single nucleotide precision. In the case of pre-tRNA splicing, there is no sequence or length conservation in the intron, but the single intron occurs at the same location in all pre-tRNAs (see Section 2.1). In addition, the two exons of all pre-tRNAs appear to be able to fold into a conserved secondary and tertiary structure very similar to that of the mature tRNA (*Figure 1*). In RNA precursors containing group I or group II introns, conserved sequence and structural elements are found at the splice sites and within the introns. The introns themselves possess most of the information and the catalytic activity necessary for their own precise removal (reviewed in ref. 4; see Sections 2.2 and 2.3). In the nuclear pre-mRNAs of *S.cerevisiae*, highly conserved splicing signals are found at the splice sites and at the site of lariat formation (67 – 69; *Figure 7*). In addition, only a few *S.cerevisiae* pre-mRNAs are interrupted, usually by a single intron of a few hundred nucleotides in length; only one pre-mRNA, MATal, is known to contain two introns, which are very short (66).

In contrast, the manner in which specificity is achieved during the splicing of metazoan nuclear pre-mRNA is unclear. The majority of metazoan pre-mRNAs contain introns, and in some cases, such as collagen pre-mRNA, as many as fifty (169,170). The introns vary widely in length, from 31 nt, in the case of a simian virus 40 late transcript (57), to 60 000 nt, in the case of an *Antennapedia* intron of *Drosophila* (58,59). Only weakly conserved splicing signals have been identified in all introns, at the splice sites (60,61) and more recently, at the site of lariat formation (62–64). Although these consensus sequences are clearly important signals for the specificity and efficiency of splicing, they do not appear sufficient to account for the specificity of splice site selection during the splicing of pre-mRNAs containing multiple and/or very long introns.

4.2 The role of conserved sequences

4.2.1 Metazoan conserved sequences

The conserved sequences at the 5′ and 3′ splice sites of mammalian pre-mRNAs are known to be important from studies of the *in vivo* and *in vitro* effects of naturally occurring and site-directed mutations (reviewed in ref. 71; 83,86,94,95,99,171,172). A variety of mutations at these sites abolish or reduce accurate splicing. Often, the same mutations activate cryptic splice sites, which also conform to the consensus sequences but are completely silent in wild-type pre-mRNA (reviewed in ref. 71). Other mutations can create new splice sites by modifying existing sequences so that they resemble the consensus sequence more closely (reviewed in ref. 71).

Some mutations at the invariant position 1 of the 5′ splice site conserved sequence abolish utilization of that site and activate nearby cryptic 5′ splice sites (reviewed in ref. 71). On the other hand, point mutations at either position 1 or position 2 of the second intron of rabbit β-globin pre-mRNA have recently been shown to allow 5′ cleavage and lariat formation, while preventing subsequent cleavage at the 3′ splice site and exon ligation (99). This effect was first observed with *S.cerevisiae* 5′ splice site mutants (see Section 4.2.2). These results suggest that recognition of a normal branch structure, including at least two nucleotides derived from the 5′ end of the intron, may be involved in cleavage at the 3′ splice site and exon ligation. Mechanistically, this could mean that a factor required for the 3′ splice site reaction must recognize a specific branchpoint sequence or structural feature. Alternatively, the aberrant branchpoint structure prevents dissociation, translocation or conformational changes of a bound factor necessary for the 5′ splice site reaction. Finally, one or more factors required for the 3′ splice site reaction may be sensitive to the sequence at the 5′ splice site, and they may be activated in response to this signal prior to the 5′ splice site reaction.

A mutation of the human β-globin 3′ splice site, which deletes the conserved pyrimidines and the AG dinucleotide, not only prevents splicing but also appears to block cleavage at the 5′ splice site and lariat formation (94,95). This observation suggests that 3′ splice site recognition is necessary for events that take place at the 5′ splice site and branchpoint sequence (see also ref. 83). Binding studies are consistent with such a model (see Section 6.5). Pre-mRNAs containing mutations that affect the AG dinucleotide but leave the polypyrimidine stretch intact are competent for the 5′ splice reaction, albeit at reduced efficiencies. However, these mutations block the 3′ splice site reaction (94,95).

No evidence for a role of other sequences within introns was obtained initially from studies of the *in vivo* expression of mutant transcription units, except for an apparent minimal intron size requirement (171). In another study, the smallest intron tested was 29 nucleotides long, and it was spliced *in vivo* at a low level (172). However, if intron sequences other than the splice sites are required for splicing, activation of putative cryptic sites upon mutation of the natural sites would not be apparent by analyzing the structure of mature mRNA, unless the selection of splice sites is affected. Furthermore, quantitative effects of such mutations may not always be obvious by analyzing the *in vivo* steady-state levels of cytoplasmic mRNA, because many splicing defects are not sufficiently severe to be detected, unless the levels of accumulated nuclear pre-mRNA are also measured (163).

The existence of a weakly conserved branchpoint sequence was noticed upon characterization of RNA intermediates generated during splicing *in vitro*, which contained an unusual structure at this site (62,86,88,89,94). The branchpoint sequence can often be deleted or mutated without an obvious effect on splicing *in vivo* and *in vitro*, except for a decrease in rate (94,173 – 175). The reaction can still take place by activation of cryptic branchpoint sequences present in the intron, without affecting the structure of the mature mRNA. In some cases, point mutations of the branchpoint adenosine result in branch formation at the mutant base *in vivo* and *in vitro*, with variable effects on the subsequent 3′ splice site reaction (100,101).

The weakly conserved branchpoint sequence does not appear to contribute to the specificity of 5′ splice site selection, since the cleaved 5′ ends of a normal 5′ splice site, of a mutationally activated 5′ splice site, and of several cryptic 5′ splice sites activated by mutation of the normal site, are all joined to the same residue in the intron (86,94). However, the selection of a branchpoint sequence and of a 3′ splice site may be interdependent events, since the two elements are only separated by a short distance.

The 5′ splice site and the branchpoint sequence on one hand, and the exon borders on the other hand, must interact, since they will be joined covalently. The above observations imply that in addition, other direct or indirect interactions take place among the three sequence elements. Such interactions may contribute to the specificity of splice site selection.

4.2.2 Conserved sequences in S.cerevisiae

In yeast, the importance of the UACUAAC conserved element was first demonstrated *in vivo* by mutational analyses (67,176,177). The yeast UACUAAC box and the branchpoint sequence in higher eukaryotes were shown subsequently to be functionally analogous (178,179). The more obvious effects of mutations in the yeast intron element can be explained by what appear to be more stringent sequence requirements, which thus usually preclude activation of cryptic branchpoint sequences. Inefficient use of a cryptic UACUAAG sequence upon mutation of the natural UACUAAC element has been observed in one case (180). More often, deletion of the natural UACUAAC box prevents splicing (67,176,177), whereas point mutations in this sequence inhibit splicing to various extents (67,96–98,182,183). The effects on splicing complex assembly *in vitro* (81,96,97) and on both 5′ and 3′ splice site reactions can be quite severe, but all mutant branchpoint sequences support some splicing in at least some contexts. The position of the branch nucleotide appears to be constant, even when the invariable A residue is mutated. When the branch nucleotide is not an A, the aberrant branchpoint can also inhibit the 3′ splice site reaction, but a very low level of splicing can sometimes be detected (97,98,183).

Mutations in the highly conserved 5′ splice site have severe effects on splicing. Mutations at positions 1 or 2 in the invariant GU dinucleotide abolish splicing (96–98). Some mutants at either position allow inefficient but precise cleavage at the 5′ splice site and lariat formation, with subsequent inhibition of the 3′ splice site reaction. Mutations at position 5 result in a low level of accurate splicing. In addition, they cause aberrant cleavages near the 5′ splice site, which may be coupled to aberrant lariat formation, but the resulting molecules are not processed further (98,182,184). A mutation at position 6 partially inhibits the 5′ splice site reaction, but gives rise to detectable mRNA (183).

Mutations in the conserved AG dinucleotide at the 3′ splice site reduce splicing efficiency. Their effect on the 5′ splice site reaction, although measurable, is less than on the 3′ splice site reaction (97,185). In contrast to the metazoan 3′ splice site consensus sequence, the *S.cerevisiae* consensus includes a single pyrimidine preceding the AG dinucleotide, although some 3′ splice sites contain multiple pyrimidines (67–69; *Figure 7*). A transversion of the conserved pyrimidine in the actin intron has no effect on splicing (185), but other pyrimidines were present nearby, and the effect of their removal in this context was not examined. A truncated rp51 pre-mRNA lacking the 3' exon, and 3' splice site, gave rise to efficient 5' cleavage and lariat formation *in vitro* (186). However, the 40 or so nucleotides downstream of the UACUAAC box retained in this truncated pre-mRNA included multiple pyrimidines. In a chimeric pre-mRNA, substitution of sequences downstream of the UACUAAC box by bacterial sequences resulted in inhibition of the 5' splice site reaction (181). Some deletions or multiple substitutions between the actin UACUAAC element and the 3' splice site had little or no effect on splicing (185), whereas several insertions in the corresponding region of a chimeric pre-mRNA had strong effects on the 3' splice site reaction, but no effect on the 5' splice site reaction (181). Insertions at this site lead to use of the first AG dinucleotide downstream of the UACUAAC box (67).

In summary, in *S.cerevisiae* the 5' splice site reaction is affected by mutations at positions 1, 2, 5 and 6 of the 5' splice site (other positions were not tested), by mutations at every position of the UACUAAC box, and weakly by mutations at the 3' AG dinucleotide. In some cases, there appears to be an effect of mutations between the UACUAAC box and the AG dinucleotide on the 5' splice site reaction, but it is difficult to assess whether or not there is a requirement for at least one pyrimidine. The 3' splice site reaction is abolished by mutations at positions 1 and 2 of the 5' splice site. This same reaction is partially inhibited by mutation of the branchpoint A, by mutations of the 3' splice site AG, and by some changes in the spacing between the UACUAAC box and the 3' splice site.

The effects of mutations in the three sequence elements are for the most part qualitatively similar in metazoans and in *S.cerevisiae*. The major differences are:

(i)   the activation of cryptic 5' splice sites that are competent for both steps of the splicing reaction, which so far has only been observed in metazoans, and may reflect the more stringent conservation of this sequence in *S.cerevisiae*;

(ii)   the more common activation of cryptic branchpoints in metazoans, which undoubtedly reflects the much weaker conservation of the branchpoint sequence in metazoans;

(iii) 3' splice site mutations have stronger effects on the 5' and 3' splice site reactions in metazoans, which may reflect the less extensive 3' splice site consensus in *S.cerevisiae*.

4.3 Possible mechanisms

In summary, the 5' and 3' splice sites, as well as the branchpoint sequence, contribute to the specificity of splicing in metazoans and in yeast. Although no sequence or structural conservation has been found elsewhere in the introns or in the exons, the three conserved sequence elements are insufficient to account for the specificity of splice site selection in metazoans because similar sequences are found at many sites within the introns and exons. In fact, many cryptic splice or branchpoint sites can be activated upon mutation of the normal sites. Additional sequence or structural features must account for the lack of utilization of such cryptic sites, and of other similar sites that are not documented cryptic sites, in normal pre-mRNAs. Mutational analyses have pointed to the existence of poorly defined signals within exons and introns, in addition to the conserved elements, which can influence the efficiency of splicing in both metazoans (187–192) and in *S.cerevisiae* (177,181,193). These signals may participate in providing an optimal overall sequence and/or structural context that favors the selection of the natural splice sites.

In addition, a mechanism must exist to prevent exon skipping, since at least some 5' and 3' splice sites from different introns are known to be compatible (99,194–196). A variety of 5' to 3', 3' to 5', and bidirectional scanning models have been proposed to account for the lack of detectable exon skipping (197). However, these models cannot by themselves account for the lack of utilization of cryptic splice sites, which are often found on both sides of the normal splice sites.

Recently, a systematic study of the relative efficiencies of utilization of different splice sites in a *cis*-competition assay revealed two principles that may contribute to the specificity of splice site selection (190). The first observation is that the sequence context of a splice site, including exon sequences, can influence its relative efficiency of utilization when it is competing in *cis* for splicing factors with another splice site. This observation seems highly relevant to the observed complete preference of normal splice sites over cryptic sites. In addition, this extended definition of a splice site can be incorporated into a more plausible scanning model that could explain the lack of exon skipping; as the splicing machinery scans the RNA in search of adjacent splice sites, it would ignore cryptic sites because their sequence context makes them weak, relative to the *cis*-competing wild-type sites. The second observation, which is also consistent with a scanning model, is that when two splice sites of equal efficiency compete in *cis*, the internal site is preferred, provided that the surrounding sequences that influence its efficiency are preserved. This observation probably reflects the unknown mechanism that prevents exon skipping. Further experiments are necessary to explain these observations in mechanistic terms.

Further understanding of the rules that govern splice site selection in nuclear pre-mRNA should be gained when the factors necessary for splicing are purified in active forms, and their interactions with normal and mutant pre-mRNAs are studied in detail. At present, it seems clear that the base-pairing interaction between the 5' terminus of U1 snRNA and 5' splice sites can contribute to the specificity of 5' splice site selection, as originally proposed (123,124; *Figure 9*). This is shown by elegant *in vivo* experiments in which a mutant 5' splice site can be partially rescued by a transfected suppressor U1 RNA gene containing a compensatory

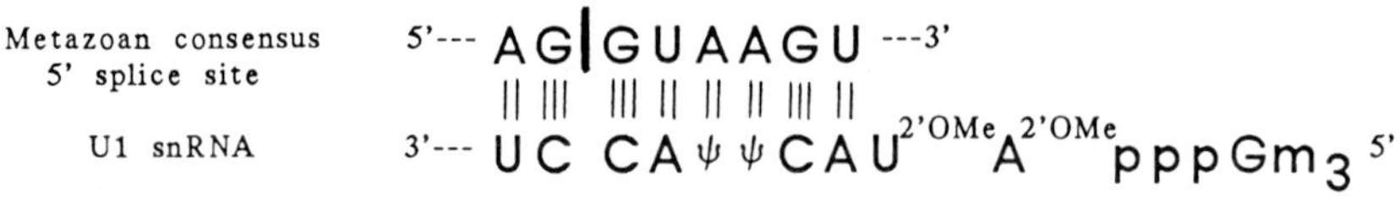

Metazoan consensus
5' splice site

U1 snRNA

$$5'--- \text{A G} | \text{G U A A G U} ---3'$$
$$|| \ ||| \ \ ||| \ || \ \ || \ \ || \ \ ||| \ \ ||$$
$$3'--- \text{U C} \ \ \text{C A} \psi \ \psi \ \text{C A U}^{2'OMe} \text{A}^{2'OMe} \text{p p p Gm}_3 \ 5'$$

Figure 9. Complementarity between U1 snRNA and the metazoan consensus 5' splice site. A model for the base-pairing interaction between the 5' terminus of U1 snRNA and the metazoan consensus 5' splice site is shown (see refs 105,123,124). Pseudouridine residues are indicated by ψ; Gm_3 and ppp denote the trimethylguanosine and triphosphate portions of the U1 snRNA cap; 2'OMe denotes ribose 2' *O*-methylations.

mutation (105). Suppression was studied in the context of the adenovirus E1A 13S and 12S 5' splice sites, which are spliced into a common 3' splice site. So far, only compensatory changes in U1 snRNA that affect the degree of complementarity at positions 3, 4 and 5 of a 5' splice site have been examined.

Efficient suppression was obtained at position 5 of the E1A 12S 5' splice site. The 12S mutation is a double point mutation at positions 5 and 6 of the 5' splice site. The former mutation decreases, and the latter increases, the degree of complementarity with the U1 snRNA 5' terminus. Nevertheless, this double mutation inactivates the 12S 5' splice site when a competing 13S 5' splice site is present in the same transcription unit. In the absence of a 13S 5' splice site, this mutated 12S site is still active. The compensatory change in U1 that restores complementarity at position 5 can reactivate the 12S 5' splice site in the presence of a wild-type 13S site. It should be noted that in this situation the suppressor U1 snRNA and the doubly-mutated 12S site have a greater degree of complementarity than the wild-type U1 snRNA and wild-type 12S site. Furthermore, the suppressor U1 snRNA reduces by one base pair the complementarity to the wild-type 13S site. Expression of the suppressor U1 snRNA may have been as high as 25% in the transfected cells. Therefore, the combined effects on the competing 12S and 13S splice sites may be responsible for efficient suppression.

A point mutation in position 3 of the 13S 5' splice site, which inactivates splicing, was not suppressed by a compensatory change in U1. However, in the context of a double 12S and 13S 5' splice mutant, a normally undetectable cryptic site can be activated by expression of a U1 snRNA with increased complementarity at position 3 of this cryptic 5' splice site. Finally, in the context of wild-type E1A, a mutant transfected U1 gene that increases complementarity at position 3 of the 12S 5' splice site, and decreases it at position 3 of the 13S site, increases utilization of the 12S site relative to the 13S site.

In summary, the extent of complementarity between 5' splice sites and the 5' terminus of U1 snRNA appears to influence 5' splice site selection.

Not all of the potential base pairs have been examined, and furthermore, the relative contribution of each base pair seems to depend on the context of its 5' splice site, as well as on the presence of competing 5' splice sites. Not all 5' splice site mutations may be suppressible in this fashion, because both the 5' splice site and the 5' terminus of U1 snRNA may have additional roles other than base-pairing with each other. For example, the G at the 5' splice site position 1 forms a 5'–2' phosphodiester bond with the branchpoint adenosine to form the lariat intermediate when the 5' splice site is cleaved (*Figure 8*). This reaction may involve sequence-specific recognition by factors other than U1 snRNA. Recognition of several bases of the 5' splice site within the lariat structure appears to influence the efficiency of the 3' splice site reaction (99). In addition, the 5' terminus of U1 snRNA may interact with other factors during the course of the splicing reaction. Finally, an additional constraint on suppression of certain 5' splice site mutations by compensatory changes in U1 snRNA may be imposed by the presence of modified nucleosides in U1 snRNA; the precise sequence at the 5' terminus may influence nucleoside modification, which could have a direct or indirect role in splicing. In the future it may be useful to study suppression *in vitro*, because some 5' splice site mutations may only be suppressed for part of the splicing reaction, leading to the formation of intermediates that may be unstable *in vivo*.

An important question is how U1 snRNP interacts with cryptic 5' splice sites. After incubation in a splicing extract, a cryptic 5' splice site activated by mutation at position 1 of the natural 5' splice site of a globin intron appears to bind U1 snRNP (198). This was determined by protection of specific RNA fragments from T1 ribonuclease in the presence of anti-(U1) RNP antibodies. The mutated site was not protected under the same conditions. A transcript containing the wild-type 5' splice site generated an analogous protection pattern but, in addition, it protected a shorter fragment; a corresponding fragment was not detected with either mutant or activated cryptic sites. These observations indicate that even though U1 snRNP may interact functionally with activated cryptic 5' splice sites, this interaction may differ from that with wild-type 5' splice sites (198). At least two kinds of interaction between U1 snRNP and normal 5' splice sites are detected by this assay (106,113,198) and it is not clear at present whether all these interactions are equally relevant to splicing and splice site selection (see Section 6.5). However, the *in vivo* suppression experiments described above argue that the extent of complementarity with U1 snRNA plays a role in the activation of a cryptic 5' splice site (105).

Sequence-specific binding of additional factors at the branchpoint sequence and the 3' splice site have been described (see Section 6.5); the U2 snRNP can bind at the branchpoint sequence (106), and a polypeptide that under some conditions may be associated with U5 snRNP can bind

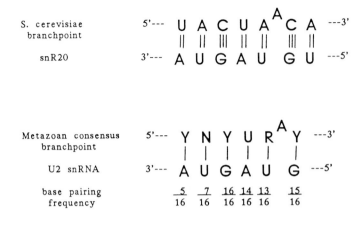

Figure 10.   Complementarity between snR20 and the UACUAAC box. A model for the base-paring interaction between the conserved U2-like domain of the *S.cerevisiae* RNA snR20 and the conserved branchpoint sequence is shown (see refs 117,119). The model has been tested by suppression of mutations at positions 3 and 4 of the UACUAAC box (119). A similar interaction has been proposed for the corresponding domain of metazoan U2 snRNA and the metazoan consensus branchpoint sequence, as shown (106,117,119). The frequency of base-pairing (including G·U base pairs) at each position for 16 natural vertebrate and viral branchpoints (see *Figure 7*) is shown.

at the 3′ splice site (109,199,200). Further studies are necessary to determine the precise nature of these interactions and their contributions to the splicing pathway and to splice site selection specificity. A large *S.cerevisiae* RNA termed snR20 or LSR1 has recently been shown to interact with the yeast branchpoint sequence by base-pairing (117,119). Compensatory mutations in this RNA can suppress splicing defects due to mutations in the conserved UACUAAC box (119). Since this interaction involves a conserved domain of snR20 that is highly homologous to a domain of metazoan U2 snRNA, a related base-pairing interaction between U2 snRNA and the weakly conserved branchpoint sequence may be involved in recognition of this site during metazoan pre-mRNA splicing (117,119; *Figure 10*).

It is possible that different specificity factors can interact with different subsets of splice sites. Alternatively, the same specificity factors may interact in different modes with different classes of splice sites. For example, many 5′ splice sites can be uniquely assigned to one of four distinct classes on the basis of sequence: AG/GUA, /GUAAGU, RG/GUGAG, AG/GUNNGU (201). If 3′ splice sites and branchpoint sequences also fall into a few distinct sequence classes, and if only some combinations are recognized productively by the splicing machinery, then a limited number of sequence-specific factors could account for a large degree of specificity.

During the course of the splicing reaction, the 5' splice site and the branchpoint sequence are brought together. At some point, the exon borders must be properly aligned. The interactions responsible for the alignment of these sequences are critical determinants of the specificity of the reaction. However, the nature of these interactions is presently unknown. They may involve RNA – RNA interactions between different sequences in the pre-mRNA, as well as RNA – RNA, RNA – protein, and/or protein – protein interactions among factors bound at specific sites on the pre-mRNA.

Complementarity between the 5' splice site and the branchpoint consensus sequences has been noted in both *S.cerevisiae* and metazoan pre-mRNAs (177,202,203). Base-pairing between these sequences could be involved in their alignment prior to the 5' splice site reaction. However, double mutations that restore this complementarity in an *S.cerevisiae* intron fail to suppress the splicing defect caused by each individual mutation (183). Lack of suppression does not rule out an interaction by RNA base-pairing, because there could be pleiotropic effects on splicing, of mutations at these essential sites. In metazoan introns, the complementarity is virtually eliminated when some cryptic branchpoints are used instead of the mutated natural branchpoint, but splicing can still take place, albeit inefficiently (175). In summary, the proposed base-pairing interaction between the 5' splice site and the branchpoint sequence does not appear to be essential, and may not exist. However, the possibility that such an interaction may influence the rate of the splicing reaction and/or the choice of splice sites has not been ruled out.

The higher order RNA and/or RNP structure of the pre-mRNA may contribute significantly to splice site selection. For example, RNA secondary and tertiary structure may position the natural splice sites in an optimal configuration (204). Extensive RNA secondary structure introduced deliberately into pre-mRNAs can cause exon skipping *in vivo* and *in vitro* (205). However, the extent of exon skipping observed with the artificial pre-mRNAs appeared to be less *in vivo* than *in vitro* (205). One possible explanation is that RNP assembly on a nascent pre-mRNA results in a different higher order structure than that obtained with a full length pre-mRNA. Alternatively, RNA unwinding activities may act to different extents *in vivo* and *in vitro*. Exon skipping was observed *in vitro* with a natural pre-mRNA, and this could also be due to aberrant RNP assembly (99). Alternatively, splice site commitment may occur during pre-mRNA synthesis *in vivo*, by binding of splicing factors on the nascent pre-mRNA (99).

In summary, it appears that in addition to the three conserved sequence elements, other, more subtle signals are present in pre-mRNA, which influence the specificity of splice site selection. The nature of these signals and their interactions with the splicing machinery are unknown. The informational content of these signals may be at the level of higher order structure, or at the level of sequence context. These signals are difficult to study systematically because they may not map to well-defined sequences, and because they do not appear to be essential for the utilization of isolated splice sites. Their effects on the rates of utilization of individual splice sites may only be detectable in *cis*-competition assays (190,206,207). Nevertheless, subtle differences in rates can be selected during the evolution of intron-containing genes, resulting in optimal sequence arrangements that may be responsible for the specificity of splice site selection by the general splicing machinery.

In some specialized cases, which are very difficult to study in a meaningful context, specialized sequence arrangements may have evolved to facilitate proper splicing. For example, pre-mRNAs containing fifty introns may contain very few cryptic splice sites, and not all their 5′ and 3′ splice sites may be compatible. Pre-mRNAs containing extremely long introns may have unusual higher order structures, they may lack cryptic splice sites positioned in favorable sequence contexts, their introns could be excised in a stepwise manner, or their splicing may be controlled by specialized factors.

5. Biochemistry of pre-mRNA splicing

5.1 Pre-mRNA splicing *in vitro*

Cell-free systems capable of efficiently and accurately splicing exogenous pre-mRNA have made it possible to analyze the biochemical requirements for pre-mRNA splicing (84–87). In addition, these systems allowed the identification of processing intermediates, which has helped the study of the reaction mechanisms (62,86,88,89). A variety of manipulations of the splicing extracts are helping to identify the numerous components of the splicing machinery (82,83,90–93,103,106–108,111,114,115,156,208). Finally, the high degree of specificity of the splicing reaction is at least partly preserved *in vitro*, thus permitting a study of the mechanisms of splice site selection (86,94–97,99,100,174,175,180,181,183,186,190).

Pre-mRNA is now commonly generated by transcription of cloned genes fused to bacteriophage promoters, with purified bacteriophage RNA polymerases (135). These *in vitro*-generated transcripts are efficiently processed *in vitro* in both mammalian and yeast systems (86,87). A salt wash of HeLa cell nuclei (209) is widely used to obtain efficient splicing extracts (86). A whole cell extract obtained from *S.cerevisiae* spheroplasts is employed to study yeast splicing *in vitro* (87). When well characterized, simple pre-mRNAs are employed, it is possible to assay splicing by direct gel analysis and autoradiography of radiolabeled RNAs (62,88).

Pre-mRNA splicing in mammalian, but not in *S.cerevisiae*, *in vitro* systems takes place after a time lag of several minutes (84–87). During this lag the pre-mRNA is assembled into fast-sedimenting RNP complexes

(82,83); similar complexes were first described in *S.cerevisiae* (81). The lag observed in HeLa extracts may reflect the association of multiple components, some of which may have become limiting during isolation, and hence this lag may or may not be physiologically significant. Another reason for the time lag observed *in vitro* may be that assembly of a suitable ribonucleoprotein structure is inherently more difficult with a full length pre-mRNA than with a nascent pre-mRNA.

The rates of splicing of different pre-mRNAs *in vivo* and *in vitro* have not been compared systematically. Pulse-chase measurements of the *in vivo* splicing rates of an adenovirus E3 pre-mRNA indicate that intermediates are found after a lag of not more than 2 min, and spliced mRNA appears after not more than 4 min (160). However, splicing of this pre-mRNA *in vitro* has not been reported. The first two leaders of the adenovirus major late pre-mRNA are spliced *in vitro* only after a lag of up to 45 min (84,85). *In vivo* the same spliced leaders are already detectable after 10 sec, whereas the splicing of the second and third leaders is preceded by a lag of about 15 min (148).

The known biochemical requirements for nuclear pre-mRNA splicing *in vitro* include monovalent and divalent cations, and ATP (84–87). Creatine phosphate is essential to allow regeneration of ATP by endogenous creatine phosphokinase. Hydrophilic polymers can stimulate the reaction by excluded volume effects (86,87). The optimal conditions may differ for different substrates, or as the components of the splicing apparatus are purified from crude extracts.

5.2 The role of ATP

The exact role or roles of ATP is an important unanswered question about the mechanism of nuclear pre-mRNA splicing. Early experiments showed that there was an ATP requirement during the initial time lag (85,86). This was demonstrated by the fact that pre-incubation of the splicing extracts with the pre-mRNA in the absence of ATP did not abolish the time lag between the addition of ATP and the first covalent changes in the pre-mRNA. It was shown later that formation of the 60S splicing complex, which occurs during the time lag, is ATP-dependent (82,83). ATP analogs containing non-hydrolyzable bonds at the α-β or at the β-γ positions do not allow the 5' splice site reaction to take place.

The involvement of ATP in spliceosome formation has not been explained mechanistically. Some of the events that take place during assembly of the spliceosome have been described, particularly with respect to the specific binding of snRNPs to the pre-mRNA (82,106,109–113; see Section 6.1). Of these interactions, binding of U2 snRNP at the branchpoint sequence, and one form of binding of U1 snRNP at the 5' splice site, require ATP (106,113,198). However, these binding steps have been studied in the crude extract, and hence it is not known if the ATP requirement is for the binding itself, or whether ATP is necessary for as yet undefined events preceding the binding of these snRNP particles. Apparent ATP-dependent conformational changes of U1, U2, and U4/U6 snRNPs have been described (90,106 – 108,112), but their significance in splicing, if any, is presently unknown.

In addition to the ATP requirement for the formation of a fast-sedimenting splicing complex with pre-mRNA (81 – 83), ATP is also required for conversion of this complex to one containing the intermediates resulting from the 5′ splice site reaction (82,93). ATP also appears to be required for the conversion of the exon 1 and lariat – exon 2 intermediates to the final products of the reaction (140; D.Frendeway, A.Kramer and W.Keller, unpublished results). Thus, ATP is required for, or prior to, the two cleavage – ligation reactions, or to maintain the structural integrity of the intermediates generated in the preceding steps. Neither α-β nor β-γ non-hydrolyzable analogs appear to be able to support these partial reactions, suggesting that hydrolysis of both types of bonds in ATP could be necessary for splicing. These experiments do not demonstrate that ATP is required as an external source of energy, since its involvement exclusively as a cofactor cannot be ruled out on the basis of inhibition studies with non-hydrolyzable analogs.

In light of the ATP requirements for the 5′ splice site and 3′ splice site cleavage – ligation reactions, it is of interest to determine whether ATP hydrolysis is directly required for the two ligation steps of the reaction: lariat formation, and exon joining. It seems likely, by analogy to the recently discovered mechanism of mitochondrial group II splicing (53,54; see Sections 2.2 and 2.3), that these are transesterification mechanisms (reviewed in ref. 4), in which the energy gained by hydrolysis of the phosphodiester bonds at the 5′ and 3′ splice sites is used to form the new phosphodiester bonds at the lariat branch and at the spliced junction. This concerted mechanism does not require an external source of energy, and hence ATP may not be directly involved in these ligation steps.

All known enzyme-catalyzed RNA ligation mechanisms require ATP. These include the mechanisms of bacteriophage T4 RNA ligase (18 – 20), and of wheat germ (16,17), yeast (14,15) and HeLa (25,26) RNA ligases. The T4, wheat germ, and yeast RNA ligases, which appear to be involved in tRNA splicing, require ATP to form a transient enzyme – AMP covalent complex that subsequently activates a 5′-phosphomonoester terminus by adenylylation (*Figure 2*). This reaction requires hydrolysis of the α-β bond of ATP. In addition, a kinase activity intrinsic to the wheat germ and yeast enzymes hydrolyzes the β-γ bond of ATP to first phosphorylate the 5′-hydroxyl terminus; this phosphate later forms the phosphodiester bond at the spliced junction. The ATP requirement for the HeLa RNA ligase is not understood (26), since the phosphodiester bond at the spliced junction is derived from the 2′,3′-cyclic phosphodiester (21,23) (*Figure 3*). The enzyme is not active on 2′- or 3′-phosphomonoester termini (26), suggesting that the ATP requirement is not due to an associated ATP-dependent RNA cyclase (25,27). Finally, the HeLa RNA ligase requires a 5'-hydroxyl terminus, and is inactive on 5'-phosphomonoester termini, and therefore, phosphorylation of the 5' terminus by a putative associated kinase activity is unlikely to occur (26). Perhaps ATP is required for activation of the 2',3'-cyclic phosphodiester by adenylylation, coupled to formation of a 3' phosphate by a putative 2'-cyclic phosphodiesterase activity, and prior to the joining of the exons.

In the case of the two ligation reactions that occur during nuclear pre-mRNA splicing, the 2',5'-phosphodiester bond at the branch is derived from the 5' phosphate of the cleaved 5' splice site, whereas the phosphodiester bond at the spliced junction is derived from the 5' phosphate at the cleaved second exon (62,89; *Figure 8*). These ligation reactions may be enzyme-catalyzed, and may involve activation of 5'-phosphate termini by adenylylation. However, no evidence for adenylylated RNA or enzyme intermediates in pre-mRNA splicing has been reported.

The magnesium requirement for the 5' splice site reaction (86) and for the 3' splice site reaction (140) does not distinguish between a trans-esterification mechanism and an enzyme-catalyzed cleavage–ligation mechanism, because all such reactions known so far require a divalent cation (4,13,14,53,54).

6. Factors involved in pre-mRNA splicing

Two types of factors have long been thought to be involved in nuclear pre-mRNA splicing. These include the nucleoplasmic snRNPs of the U family (reviewed in refs 5,104) and the polypeptides that associate with hnRNA to form hnRNP (reviewed in refs 6,155). With the development of *in vitro* splicing systems, the involvement of some of these factors has been tested directly. In addition, a number of activities necessary for splicing *in vitro* have been identified biochemically (86,91,92,102,103) but the physical identity of these components is presently unknown. In *S.cerevisiae* several of the *rna* genes are necessary for splicing. Extracts prepared from strains bearing conditional alleles of several *rna* genes are being used to dissect the function of the corresponding gene products in pre-mRNA splicing (93,114).

6.1 Small nuclear ribonucleoprotein particles

Small nuclear RNAs are found in all eukaryotic organisms so far examined, including metazoans, fungi, protozoans and plants (compiled in ref. 210). These RNAs are highly conserved and are usually associated with multiple polypeptides. Several members of the U family of snRNPs have been shown to be involved in splicing in mammalian cells. The evolutionary conservation of the U snRNA sequences (210) suggests that these particles are involved in splicing in other eukaryotes as well. Small RNAs in *S.cerevisiae* are significantly different from those of higher eukaryotes (118). Partial sequence homology can be discerned in some cases, however, and recent experiments show that several *S.cerevisiae* small RNAs are involved in splicing.

6.1.1 Mammalian snRNPs

The abundance of snRNP particles in higher eukaryotes permits the easy detection of their snRNA moieties, and these particles have been known for a long time (reviewed in refs 5,104). In fact, the ease of isolation of the snRNA moieties by preparative gel electrophoresis permitted both the direct determination of their RNA sequences, and the cloning of the genes encoding them. The associated polypeptides have been identified by immunoprecipitation with antibodies that recognize snRNP determinants. Recently, cDNAs encoding some of these polypeptides have been cloned (211,212). Some physical properties of the snRNAs (compiled in refs 210,213), and the names and apparent molecular weights of the associated polypeptides (214–217), are described below.

The U-class of snRNPs includes the major species U1 (164 nt), U2 (187 nt), U4 (141 nt), U5 (117 nt) and U6 (106 nt), which are nucleo-plasmic, and estimated to be present in 200 000 to 1 000 000 copies per cell (218). U3 (217 nt), and possibly the recently identified U8 (140 nt), are localized in the nucleolus. The subnuclear localization of U7 (56 nt), and of the recently identified U9 (130 nt) and U10 (65 nt), are unknown. All these snRNAs contain a trimethylguanosine cap, with the exception of U6 snRNA which contains an unidentified cap structure although it is an RNA polymerase III transcript (219,220). The primary sequences of many snRNAs derived from different species are known (compiled in ref. 210). Modified nucleosides, including pseudouridine, N6-methyl-adenosine, 2'O-methylated A, G, C, and U, and 2-methylguanosine, are found in some of these RNAs. All known mammalian snRNAs, except U3, are associated with the Sm antigen, which is recognized by a variety of autoimmune antisera. U4 and U6 snRNAs exist predominantly or exclusively as a single snRNP particle (221,222).

The major nucleoplasmic snRNPs U1, U2, U4/U6, and U5 share six major polypeptides, known as B (28 kd), B' (29 kd), D (16 kd), E (12 kd), F (11 kd), and G (9 kd), and a minor polypeptide D' (15.5 kd). Polypeptides A (34 kd), C (22 kd), and an unnamed polypeptide of 68–70 kd are unique to U1 snRNP; polypeptides A' (33 kd) and B" (28.5 kd) are unique to U2 snRNP (217 and references therein).

When snRNA sequences became available, it was proposed that the U1 snRNP might serve to align splice sites by virtue of the complementarity between two adjacent sequences in U1 snRNA and the consensus 5' and 3' splice sites (123,124). Phylogenetic analysis of U1 sequence conservation was consistent with a possible involvement of the

5′ terminus of the snRNA, which is complementary to 5′ splice sites, but not with the involvement of an adjacent sequence, which in some but not all species bears weak complementarity to 3′ splice sites (223). As shown in *Figure 9*, the complementarity between the consensus 5′ splice site sequence in higher eukaryotes and the relevant sequence near the 5′ terminus of U1 snRNA involves eight contiguous base pairs, including three G-C base pairs, three A-U base pairs and two A-ψ base pairs (pseudouracil can base-pair with adenine; see for example ref. 224). Ignoring the post-transcriptional modifications, this octamer sequence should occur on the average once every 66 kb. This striking complementarity is consistent with the notion that the U1 snRNA may be responsible for 5′ splice site recognition. It is not necessary to make any assumptions about the stability of the base-paired sequences, as the energy contribution may well be affected by polypeptide–RNA interactions (225). The complementarity is of course less extensive with most 5′ splice sites than with the consensus sequence, since the former often deviate from the consensus at several positions (*Figure 7*).

In one test of the model, the U1 snRNP was extensively purified by ion-exchange, gel filtration and hydrophobic interaction chromatography, and shown to bind to the 5′ splice site of a mouse β-globin pre-mRNA synthesized by T7 RNA polymerase (225). However, it is known that the 5′ terminus of the U1 snRNA is exposed in intact snRNP particles (226), and hence this interaction could be due to RNA–RNA base-pairing under conditions that are not necessarily relevant to splicing. It was shown, however, that the snRNP polypeptides appeared to be necessary for the specificity of binding (225).

To determine whether one or more snRNPs are involved in splicing, inhibition studies were carried out with antibodies specific for snRNPs. First, it was shown that anti-Sm antisera could inhibit splicing of adenovirus major late pre-mRNA in isolated nuclei (227). Similarly, several monoclonal and polyclonal anti-snRNP antibody preparations were shown to inhibit splicing in an *in vitro* system (228). This study suggested that U1, but not U2 snRNP was involved in splicing. However, these early studies could not be interpreted unambiguously, because of the possibility that the inhibition of splicing is caused by protease activity in some samples of serum, ascites fluid, and serum-supplemented tissue culture medium. To overcome this problem, it is desirable to use purified antibodies, and to show that activity can be restored by addition of purified snRNPs. Anti-Sm polyclonal antibodies purified and immobilized on protein A-Sepharose have been used to partially deplete extracts of their snRNPs, and concomitantly, of their splicing activity (208), but complementation of extracts depleted in this manner with snRNP fractions has not been reported.

Splicing can be completely inhibited by treatment of the extracts with micrococcal nuclease, followed by inactivation of the nuclease with EGTA

(90). It is possible to complement the nuclease-treated extracts for splicing with chromatographic fractions containing all the snRNPs (90). However, snRNPs have not yet been extensively purified in a demonstrably active form by any method, and therefore the total number of snRNPs involved in splicing is still unknown. Deproteinized HeLa RNA cannot restore activity to a microccocal nuclease-treated extract, suggesting that snRNP activity also requires the associated polypeptides, and that reassembly of snRNPs does not occur efficiently under these conditions.

Site-directed RNA cleavage has been used to ascertain whether snRNAs whose sequence is known are involved in splicing, or in other reactions (208). Sequences in the snRNAs that are accessible within the native snRNP particles are digested in a site-specific manner by hybridization to complementary oligonucleotides and incubation with RNase H. This enzyme is specific for RNA – DNA hybrids and cleaves the RNA strand. Using this method, it was shown that removal of eight 5′ terminal nucleotides plus the trimethylguanosine cap of U1 snRNA in splicing extracts abolished their splicing activity (208). In this study, only half of the U1 snRNP present in the extracts was digested, but this treatment was apparently sufficient to abolish splicing of an adenovirus major late pre-mRNA.

To rule out artifactual cleavage of the pre-mRNA by RNase H, which requires as few as four base pairs of continuous or hyphenated complementarity (229), splicing can be assayed by direct analysis of relatively short pre-mRNAs, so that the integrity of the pre-mRNA can be simultaneously assessed (90,106). Furthermore, the oligonucleotides used for site-directed cleavage of snRNAs in splicing extracts can be destroyed by DNase I treatment prior to addition of the pre-mRNA (90). Using this approach, the apparent involvement of the 5′ termini of U1 as well as of U2 snRNPs in the splicing reaction, in particular in the 5′ splice site reaction, was reported (90,106). In these experiments, substantial inhibition of splicing could be accomplished only when almost all the snRNA was digested, suggesting that both snRNPs are present in large excess over other splicing components. Interestingly, inactivation of splicing, and extensive digestion of the snRNAs, required prior incubation of the extracts in the presence of ATP, suggesting that the accessibility of the snRNA 5′ termini was increased as a result of an ATP-induced conformational change of the snRNP particles.

Recently, the same methodology was employed to demonstrate the involvement of the U4/U6 snRNP complex in the first step of the splicing reaction (107,108). Both U4 and U6 snRNAs appear to be necessary for splicing, as judged by this assay. Conditions for digestion of U5 snRNA in native particles using several oligonucleotides could not be found (108), and hence the requirement of U5 snRNP for splicing could not be similarly tested. However, a possible role for this particle is suggested by specific RNA binding experiments, which are described below (109).

As mentioned above, the ability of the 5' terminus of U1 snRNA to interact with 5' splice sites by base-pairing in a functionally significant manner has recently been demonstrated by the development of an *in vivo* expression system for suppressor U1 snRNAs. Compensatory mutations in a transfected U1 gene can sometimes restore utilization of mutant 5' splice sites *in vivo* (105; see Section 4.3).

In summary, it appears that all the major nucleoplasmic snRNPs, i.e. U1, U2, U4/U6 and U5 are involved in splicing. The first three particles are known to be necessary for, or prior to, the 5' splice site reaction, but they could also participate in later events. U1 snRNP is involved in 5' splice site recognition, which involves RNA – RNA base-pairing, and perhaps RNA – protein interactions. The mode of action of the other snRNPs is less certain. Base-pairing models for binding of U2 snRNP at exon borders (230) or at the branchpoint (64) have been formulated, but the complementarities are not very compelling, and these models have not been tested experimentally. A different base-pairing model between U2 snRNA and the branchpoint has been proposed by analogy to the interaction between snR20 and the UACUAAC box in *S.cerevisiae* (117,119; *Figure 10*). RNA-binding experiments show that U2 snRNP is capable of binding selectively at branchpoint sequences in a manner that is sensitive to mutations that affect splicing (106; see below). An snRNP tentatively identified as U5 can bind at 3' splice sites, also in a manner that is sensitive to splicing mutations (109; see below). However, the nature of these interactions is unknown, and their precise relevance to splicing and splice site selection remains to be determined. No binding sites for U4/U6 snRNP have been identified. The possible involvement of minor snRNP species (210,213) in splicing has not been addressed. Of these minor species, only U7 snRNA has been assigned a role: it participates in 3' end formation of histone pre-mRNAs (reviewed in ref. 144; Chapter 4).

Finally, although models of snRNP involvement have emphasized recognition of sequence elements by base-pairing, snRNP polypeptides may play important structural or enzymatic roles. Furthermore, the discovery of RNA catalysis (reviewed in refs 4,30,31) sets a precedent for a possible catalytic role of the snRNA constituents in splicing. Further definition of the roles of snRNPs in splicing will require their purification in active forms.

Purified snRNPs have been obtained by procedures that appear to maintain the structural integrity of the particles, as judged by the presence of the snRNA and polypeptide moieties, and by immunoprecipitation with anti-Sm or anti-U1 antibodies (214,231,232). However, in the absence of functional assays, it has not been determined whether these purified particles have maintained their activities. Filter-binding assays have also been used for partial purification of U1 snRNP (233,234). Given the involvement of snRNPs in pre-mRNA splicing, it is of particular interest to determine whether these particles have any enzymatic activities, or whether they function solely in splice site recognition.

Antibodies specific for the $N2,N2,7$-trimethylguanosine (m_3G) cap present at the 5′ ends of snRNAs U1 through U5 (109,235,236) have been shown to be useful for the purification of intact U1, U2, U4/U6 and U5 snRNPs (215,217,221). These particles are gently recovered by affinity elution with free m_3G nucleoside. Although this is the most powerful, and at the same time probably the gentlest, procedure available for snRNP purification, it remains to be shown that the purified particles are active, particularly in pre-mRNA splicing. It is possible that only a sub-population of each snRNP is active in splicing, and these sub-populations may not be accessible to anti-m_3G antibodies, perhaps due to an m_3G cap binding protein. In addition, special conditions may be necessary to preserve the activity of these particles throughout the purification. Separation of immunoaffinity purified m_3G-containing snRNPs from each other can be accomplished by a combination of immunoaffinity chromatography with anti-$N6$-methyladenosine antibodies, and ion-exchange chromatography (217).

6.1.2 Yeast snRNAs

Until recently, S.cerevisiae was not thought to possess snRNPs. Attempts to identify U snRNA homologs by nucleic acid hybridization, or by immunoprecipitation with anti-Sm antisera, yielded negative results (116,123). Eventually, a large number of metabolically stable small RNAs containing a 5′ trimethyl cap were identified (116,118). In contrast to metazoan snRNAs, the yeast snRNAs are extremely rare, with an estimated abundance of 10–500 copies per cell. They are encoded by single copy genes, and they range in size from 120 to 1200 nucleotides. Whether all these small RNAs are associated with polypeptides is not certain at present. Some of these RNAs can associate with a heterologous Sm antigen upon microinjection into frog oocytes (118,237). Recently, the presence of Sm and U1 (RNP) determinants in native snRNPs from S.cerevisiae and other fungi was reported (237). In general S.cerevisiae snRNAs appear to be substantially different from the snRNAs of other yeasts and fungi, whereas the latter more closely resemble metazoan snRNAs (117,118,237).

Pre-mRNA splicing in a yeast in vitro system is sensitive to micrococcal nuclease treatment of the extracts (92), and two out of three partially purified fractions necessary for splicing contain several small RNAs. However, considering that at least 24 small RNAs have been identified in S.cerevisiae and given their low abundance, it is difficult to examine systematically their involvement in splicing solely by biochemical fractionation techniques. Gene disruption techniques have been used to identify essential small RNAs. snR3, 4, 5, 8, 9 and 10 are not essential for growth (238 and references therein), whereas disruptions of snR7, 14, and 20 have a lethal phenotype (117,238).

snR7$_s$ (178 nt), snR7$_1$ (213 nt), snR14 (163 nt) and snR20 (1175 nt) (117,118) are associated with splicing complexes (115; see Section 6.4). snR20 (LSR1) is a large trimethyl-capped RNA that contains strong homology to metazoan U2 snRNA, and weaker homology to U4, U5 and U6 snRNAs (117). Recent genetic suppression experiments identified compensatory mutations in snR20 that can suppress splicing mutations in the UACUAAC sequence (119). These results demonstrate a functionally significant base-pairing interaction between the U2-like domain of snR20 and the branchpoint sequence (*Figure 10*). Since human U2 snRNP is capable of binding mammalian pre-mRNA at the branchpoint sequence (106), an analogous base-pairing interaction may be important for metazoan pre-mRNA splicing (117,119; see *Figure 10*). The weaker conservation of the metazoan branchpoint sequence may be compensated by interactions of a different nature. For example, U2 binding could be stabilized by interactions with components bound at the 3' splice site (109,199,200).

6.2 Protein factors

In addition to the snRNPs, a number of protein factors appear to be required for splicing, a reaction that is sensitive to protease treatment (84). These factors have been identified as activities that are not sensitive to micrococcal nuclease, and are sensitive to sulfhydryl reagents, and/or to heat, and/or are chromatographically separable from any detectable RNA species (90,91,102,103). None of these observations can rule out the possibility that many or all of these factors are constituents of snRNPs that can be dissociated and reassembled under some conditions (90). Four activities necessary for splicing were identified that did not correlate with the presence of intact snRNAs, and were named SF2 (splicing factor 2), SF3, SF4A, and SF4B (90). Two similar activities were identified by Perkins *et al.*, termed Ia and Ib (91), that are likely to correspond, respectively, to SF4A and SF4B. An activity termed SF2 has been extensively purified (103), but based on the reported properties, this SF2 is almost certainly different from the above SF2.

The activities termed SF4A and SF3 are only necessary for the 3' splice site reaction, as is the case for fraction IA, which may be the same as SF4A. The remaining activities, as well as fractions containing snRNPs are required for or prior to the 5' splice site reaction, and may or may not be involved in subsequent steps.

None of the above protein factors has been sufficiently purified to allow assignment to a specific polypeptide. Some activities may only have a trivial involvement in the *in vitro* splicing reaction. This would be expected, for example, of protease, phosphatase and RNase inhibitors, and of creatine phosphokinase. The requirement for such activities would be expected to disappear upon further purification.

In *S.cerevisiae* three fractions have been obtained. Fractions I and II are necessary for the 5′ splice site reaction, whereas fraction III is only necessary for the 3′ splice site reaction (92). Fractions I and II contain micrococcal nuclease-sensitive components. Recently, the splicing functions of the *rna* gene products have been examined *in vitro* (93,114). The *RNA* genes were originally identified as temperature-sensitive alleles of genes involved in ribosome biosynthesis. Many of these mutations exert their effect on ribosome biosynthesis in an indirect manner, by preventing the splicing of nuclear pre-mRNAs encoding ribosomal proteins (see ref. 114 and references therein). The products of the *RNA*2, 3, 4, 5, 7, 8 and 11 genes appear to be necessary for splicing, because extracts prepared from strains bearing the corresponding temperature-sensitive mutations are heat-sensitive for splicing *in vitro* (114). Four of these genes are known to encode proteins. These gene products are all present in fractions I. All of these *rna* gene products are necessary for or prior to the 5′ splice site reaction (114), and all except the *rna*2 product are necessary for spliceosome assembly (93; see Section 6.4.2). An additional activity termed b_n has been inferred to be involved in the 5′ splice site reaction, whereas activities known as c_{III} and c_n have been inferred to be involved in the 3′ splice site reaction (93).

6.3 HnRNP proteins

Among possible candidates for protein factors necessary for splicing, which may or may not be distinct from the activities described above, are the major protein components of hnRNP particles (reviewed in ref. 6; see Section 3.3). An important question is whether hnRNP proteins exhibit any sequence-specific or structure-specific association with RNA (239). This type of interaction could conceivably position relevant sequences in an optimal three-dimensional configuration for recognition by splicing factors. A simpler role for hnRNP proteins would be to untangle long RNAs to make important sequences accessible to the RNA processing machinery. Finally, the hnRNP proteins may have enzymatic activities necessary for splicing.

The availability of *in vitro* splicing systems, and of antisera and monoclonal antibodies against hnRNP proteins, makes it possible to address directly the question of whether these proteins are necessary for splicing (156,157). Depletion of the RNA-binding C1 and C2 polypeptides from a splicing extract by immunoaffinity chromatography inhibits splicing (156). In particular, this treatment prevents the formation of splicing complexes (156). It remains to be shown whether addition of C proteins to be depleted extract restores splicing; until then, it remains possible that removal of an associated factor(s) rather than of the C proteins themselves, is responsible for the observed inhibition of splicing. No involvement of the 68 kd and 120 kd hnRNP-associated polypeptides in splicing could be demonstrated by inhibition studies with purified monoclonal antibodies (156).

6.4 Splicing complexes

6.4.1 Mammalian spliceosomes

During the time lag that precedes *in vitro* splicing in mammalian extracts (84 – 86), the pre-mRNA associates with multiple factors to form large complexes that have been analyzed by glycerol or sucrose gradient sedimentation (82,83,91,110,113,240), and more recently by equilibrium density sedimentation (241), gel electrophoresis (112) and affinity chromatography (111).

These complexes were first identified by incubating radiolabeled pre-mRNA with crude splicing extracts, and sedimenting the mixture through sucrose or glycerol gradients (81 – 83). Fractions are analyzed by measuring their content of radiolabel and the structures of the radiolabeled RNAs present in each fraction are determined by polyacrylamide gel electrophoresis. Two or sometimes three peaks of radioactivity are obtained. The largest complex sediments at 50 – 60S, although its sedimentation rate decreases with increasing salt concentrations or in the presence of heparin (82,110 – 112). These treatments probably remove both non-specific and specific components. This 60S peak contains both pre-mRNA and the splicing intermediates, and is usually termed the spliceosome, because the biochemical requirements for its formation generally coincide with those necessary for splicing.

Slower-sedimenting peaks are detected under some conditions at 22S and 35S (83), or at 40S (82,110). Based on reported properties other than the sedimentation values, the 40S and the 22S peaks are analogous, whereas the 35S peak was only detected in one study. These and other differences are probably due to different experimental conditions, primarily with respect to the substrates employed, the ionic conditions during sedimentation, and the time points examined. For simplicity, the subsequent terminology of Konarska is employed (112). The 35S-like complex is termed A, and the 55S-like complex is termed B. Similarly, the 22S- (or 40S-) like complex is termed C.

The C complex forms instantaneously, even on ice and in the absence of ATP or magnesium, which are essential for splicing. Formation of this complex does not appear to require specific RNA sequences, since a C-like complex can form with RNAs of bacterial origin. It appears to be a short-lived complex. Depletion of snRNPs from the extract does not prevent C complex formation.

The A complex sometimes but not always appears to be short-lived.

At low pre-mRNA concentrations it can form in the absence of added ATP (83). It does not form in extracts that have been depleted of snRNPs. Formation of this complex requires the presence of an intact poly-pyrimidine stretch at the 3' splice site, and is somewhat improved in the presence of the adjacent AG dinucleotide. A substrate lacking the 3' splice site and adjacent exon was reported to form a partial 50S complex (82); in a different study, a substrate lacking the 3' splice site could only form the non-specific C complex (83). This type of mutation blocks cleavage at the 5' splice site and lariat formation (94,95). In contrast, a substrate lacking the 5' splice site can still form a complex with pre-mRNA that co-sediments with the A complex (83). In this case, no cleavage – ligation reactions take place.

The B complex, known as the spliceosome, accumulates later in the reaction in an ATP-dependent manner. It contains pre-mRNA, and the cleaved exon 1 and lariat exon 2 intermediates are part of either the same complex or of a co-sedimenting complex. The final products of the re-action appear to be released from this complex, because they have lower sedimentation values.

Kinetic experiments, as well as the effects of substrate mutations, of snRNP removal, and of biochemical conditions, suggest, but do not prove, that the above complexes bear a precursor – product relationship in the order C – A – B. The pre-mRNA in the B complex can be chased into spliced products with accelerated kinetics, provided fresh extract and ATP are added (82). This experiment suggests that the B complex with pre-mRNA is a true intermediate in the reaction, although it lacks one or more factors necessary for further processing. These factors may normally interact with the B complex later in the reaction, or they may have dissociated as a consequence of the sedimentation conditions.

SnRNPs are present throughout the gradient, but at least U1 snRNP appears to be stably associated with the pre-mRNA, intermediates, and products of splicing, because these RNA species can be immuno-precipitated from the B complex and from other parts of the gradient by anti-U1 and by anti-Sm antisera (82,110,113).

Electrophoretic analysis of heparin-resistant RNP complexes from splicing reactions resolves three complexes analogous to the A, B and C complexes (112). However, the A-like complex appears to require ATP for its formation. Blotting experiments suggest that the A and B complexes analyzed in this manner contain U2 snRNA but not U1 snRNA. A similar conclusion was obtained when heparin-resistant complexes formed with biotinylated pre-mRNA were resolved by sedimentation, and were further purified by affinity chromatography with immobilized streptavidin (111). The apparent absence of U1 snRNA is inconsistent with the above immunoprecipitation results, but it may simply be due to dissociation of U1 snRNP from splicing complexes in the presence of heparin or during electrophoresis. Alternatively, the U1 snRNA may be only transiently associated with the splicing complexes, but a U1 snRNP-specific poly-peptide recognized by anti-U1 antisera is retained. U2 snRNA was found stably associated with the A-like and B-like complexes. The B-like complex was also stably associated with U4/U6 and U5 snRNPs. An additional snRNA of approximately 80 nt was also detected in association with the B-like complex. This RNA species may represent a low abundance snRNP, or it may be a 5′ terminal fragment of a major snRNA other than U1 (111).

6.4.2 Yeast spliceosomes

Similar kinds of splicing complexes have been identified using *S.cerevisiae* *in vitro* systems. These complexes have been analyzed by sedimentation (81,93,96,97), by gel electrophoresis (115,242) and by poly(A) − oligo dT affinity chromatography (242).

The splicing reaction is very rapid in the yeast *in vitro* system, and hence the products of the reaction appear with no detectable time lag (96). A 15 − 30S C-like complex and a 40S B-like complex were originally identified by sedimentation analysis (81). As in mammalian systems, formation of the B complex required ATP. Point mutations in the highly conserved branchpoint sequence, which abolish 5′ cleavage, also abolished formation of the B complex (96,97). A functional 5′ splice site is also required for formation of the B-like complex (97,115).

Electrophoretic analysis of splicing complexes that are resistant to 0.4 M salt, 20 mM EDTA and competitor RNA allows resolution into three core complexes (115). The C-like complex is no longer detected under these conditions. These spliceosome cores were termed I, II and III, in order of increasing mobility. A putative precursor − product relationship based on the kinetics of appearance and disappearance of these core complexes was proposed. In addition, snRNAs associated with these core complexes were identified by recovering the complexes from the gels. The snRNAs present in these complexes were 3′ end-labeled with pCp and RNA ligase, enriched by immunoprecipitation with anti-trimethyl cap antibodies, and analyzed by denaturing polyacrylamide gel electrophoresis. Several snRNA species were enriched compared to control reactions lacking added pre-mRNA (115). The initial core complex III contains pre-mRNA and the 1.2 kb cellular RNA snR20 (117,118). Core complex I also appears at early times, is short-lived and contains primarily pre-mRNA. It is associated with snR20, as well as with $snR7_s$, $snR7_1$ and snR14. Core complex II appears later, contains pre-mRNA, intermediates and products of splicing, and is associated with snR20, $snR7_s$ and $snR7_1$. Pre-mRNA bearing a point mutation at the branchpoint adenosine, which inhibits the 5′ splice site reaction, forms a single core complex with the same mobility as that of core complex III. Whether this core complex is still associated with snR20 remains to be tested. Whether these snRNAs interact with the pre-mRNA at specific sites, by analogy to mammalian systems, was not determined in these experiments. However, snR20 has since then been shown to interact with the UACUAAC sequence by base-pairing (119; see Section 6.1.2).

The products of the $rna2$, 3, 4, 5, 7, 8 and 11 genes of $S.cerevisiae$ appear to be necessary for splicing, since extracts derived from appropriate temperature-sensitive strains are heat-sensitive for splicing in $vitro$ (114; see Section 6.2). The corresponding heat-inactivated extracts have been examined for their ability to form splicing complexes (93). No 40S B-like complex was obtained in the absence of intact $rna3$, 4, 5, 7, 8 and 11 gene products. Therefore, these gene products are essential for one or more of the steps leading to the formation of the B complex. In addition, these gene products may or may not be stably associated with the B complex.

A 40S B-like complex with uncleaved pre-mRNA was obtained when heat-inactivated $rna2$ extracts were used (93). This complex, termed $rna2\Delta$ spliceosome, contained only pre-mRNA, indicating that the $rna2$ gene product is necessary for the 5′ splice site reaction (93,114). Formation of the $rna2\Delta$ spliceosome required ATP and an intact 5′ splice site. The $rna2\Delta$ spliceosome appears to be a $bona$ $fide$ intermediate in the splicing pathway, since following its isolation by glycerol gradient sedimentation, it can be chased into final spliced products. This chasing reaction requires wild-type $rna2$ gene product, and at least two activities termed b_n and c_n. The latter activity is necessary for the 3′ cleavage reaction. b_n and c_n are probably extrinsic factors that are not stably associated with the $rna2\Delta$ spliceosome.

6.4.3 The role of the spliceosome

Further work is necessary to identify and determine the function of the components of the mammalian and yeast spliceosomes that are necessary for the respective splicing reactions. The relationship between splicing complexes and hnRNP is presently obscure, although the hnRNP C1 and C2 proteins appear to be components of the spliceosome (156). Although spliceosomes and hnRNP cores have similar sedimentation values, intact hnRNP particles have sedimentation values of 50–300S (reviewed in ref. 6). Systematic studies to determine the sizes of splicing complexes formed with pre-mRNAs of varying lengths and multiple introns have not been conducted. HnRNP studies are generally done with heterogeneous RNA populations, whereas splicing complex studies employ discrete pre-mRNAs. Reconstitution of dissociated hnRNP particles in $vitro$ has been accomplished, and requires neither ATP nor specific RNA sequences (reviewed in ref. 6). Therefore, a relationship may exist between the C-like non-specific complex observed in spliceosome studies, and the classic hnRNP monoparticle structure.

An important question to be addressed is the status of the components of the spliceosome prior to the addition of pre-mRNA to the splicing extracts. For example, a complex may already exist with endogenous hnRNA, the components of which may be recycled in an ATP-dependent manner (90,112). Perhaps the extracts contain a preformed, ribosome-like particle consisting of several snRNPs and protein factors, which recognizes the pre-mRNA and which, together with separate factors, catalyzes the splicing reaction. However, if such complex particles exist, complementation experiments argue that many of their constituents are exchangeable (90). In a sense, the term spliceosome implies a preformed particle, which may or may not exist. Instead, individual snRNPs and other factors may sequentially assemble on, or interact with, the pre-mRNA substrate to form the splicing complexes, by analogy with transcription complexes. On the other hand, each snRNP may be thought of as a small subunit of a large complex, with each subunit recycling in the free state at the completion of each round of processing.

Some components of the spliceosome may play a structural role, by unfolding the pre-mRNA or perhaps assembling it into a precise conformation. Other components may function in the recognition of splice sites and the branchpoint sequence, and in aligning these elements prior to the cleavage and ligation reactions. The same or different components may contain active sites that catalyze these reactions. These and other questions may be answered by determining the structure and composition of the spliceosome, and studying how these parameters vary in response to pre-mRNA size, number of introns, and splicing mutations.

6.5 Site-specific binding of factors to pre-mRNA

Although the structure of splicing complexes and the identity of many of their components are presently unknown, substantial progress has been made in identifying factors that appear to bind specifically to regions of the pre-mRNA that are important for splicing.

Specific binding sites for snRNPs in pre-mRNA, as well as in the intermediates and products generated during the course of splicing *in vitro*, have been identified by immunoprecipitation of labeled pre-mRNA fragments, resulting from ribonuclease T1 digestion of splicing complexes, with anti-snRNP antibodies (106,109). Four kinds of antibodies were used for these studies: anti-Sm and anti-m$_3$G (which recognize respectively the Sm antigen and the trimethyl cap present in U1, U2, U4/U6 and U5 snRNPs), and anti-U1 and anti-U2 antisera (which are specific for polypeptides unique to U1 and U2 snRNPs, respectively). These studies identified a U1 binding site at the 5' splice site, and a U2 binding site at the branchpoint sequence. In addition, binding at the 3' splice site was due to a different snRNP that was inferred to be U5. The latter assignment was made on the basis of the following criteria: a 3' splice site fragment was only immunoprecipitated with anti-Sm and anti-m$_3$G antibodies, but not with anti-U1 or anti-U2 antisera; the binding was not affected by pre-digestion of the extracts with very high concentrations of micrococcal nuclease, and U5 snRNP was the only major snRNP that remained immunoprecipable by these antibodies after such treatment (109). The binding of these snRNPs to pre-mRNA may be direct, or it may be mediated by additional factors. Subsequent experiments suggest that the binding of U5 snRNP may be mediated by an associated polypeptide (199,200).

In this type of experiment, both the RNA substrate and the snRNAs are sensitive to digestion by the enzymatic probes. Protection of a pre-mRNA fragment by an snRNP can only be detected in this manner if the regions of the pre-mRNA that are not bound by the snRNP are more sensitive to ribonuclease than the regions of the snRNA that must remain intact to allow stable binding. Furthermore, the protected region must remain stably bound once the remainder of the pre-mRNA is digested. The antibodies are added at the same time as the ribonuclease and may protect the snRNPs from digestion. However, antibody binding may be mutually exclusive with a subset of the possible snRNP – pre-mRNA interactions. In summary, it seems likely that this method of probing snRNP – pre-mRNA interactions can only detect a subset of the relevant specific interactions. In addition, as a result of the above experimental difficulties, only a minor molar proportion of the RNA, of the order of 1%, is protected (106). Hence, it is difficult to be certain that the stable interactions that are detected are functionally significant, i.e. necessary for splicing. Fortunately, in some cases, the functional significance of these interactions can be inferred from the observation that mutations that affect splicing also result in altered binding.

The apparent interaction between U5 snRNP and the 3′ splice site occurs very rapidly, does not require ATP and is affected by mutations at this site that affect splicing, but does not require any other pre-mRNA sequences (109). The binding of a polypeptide that appears to be associated with U5 snRNP under some conditions, has essentially these same properties (199,200). The association between U2 snRNP and the branchpoint sequence occurs later, requires ATP (106), is sensitive to mutations at the 3′ splice site, and persists after lariat formation (113). Site-specific cleavage of U2 snRNA, which inhibits splicing, eliminates protection at the branchpoint sequence, but has no effect on 3′ splice site protection (113). Although the nature of the interactions between U2 snRNP and the branchpoint sequence are unknown, several base-pairing models have been proposed (64,106,117,119). One of these models is supported by the fact that an analogous base-pairing interaction between a U2-like domain of the *S.cerevisiae* snR20 RNA and the UACUAAC box has been shown to be important for splicing (119; *Figure 10*).

Several modes of U1 snRNP binding to 5′ splice sites have been described (106,113,198). Ribonuclease T1 protection experiments performed with purified U1 snRNP bound to immobilized antibodies result in protection of a 5′ splice site fragment. Binding was also detected with a mutated 5′ splice site, but not with a cryptic splice site. In contrast, binding experiments performed in a splicing extract detect extremely rapid ATP-independent binding to a wild-type 5′ splice site, but not to mutant or cryptic 5′ splice sites (type 1 binding). Upon further incubation under splicing conditions, extended 5′ splice site protection could be detected with wild-type and activated cryptic 5′ splice sites, but not with the mutant, defective 5′ splice site (type 2 binding). Because of the low recovery of protected fragments, it is not possible to tell whether the same pre-mRNA molecule gives rise to both protection patterns in sequential fashion. Hence, there may or may not be a precusor – product relationship between these two binding modes.

These experiments suggest that specific, physiologically significant binding of U1 snRNP to 5′ splice sites may require additional cellular factors present in the splicing extract but not in purified U1 snRNP. Type 2 binding is detected with RNA present in splicing complexes, and hence may reflect structural changes immediately preceding the 5′ splice site reaction (113). Type 1 binding of U1 snRNP, although seemingly predominant by the above assay, does not correlate well with function, because it is not detected with an activated cryptic 5′ splice site. This could be because the interaction is weaker and more transient than with a normal 5′ splice site (198). Alternatively, this early protection reflects a base-pairing interaction that is not necessary for splicing. The relevance of the complementarity between U1 snRNA and 5′ splice sites, established by *in vivo* suppression experiments (105), does not indicate whether only one, or both, of the interactions detected by the *in vitro* assays are important for splicing.

Although type 1 binding of U1 snRNP to a 5′ splice site can still occur with pre-mRNA bearing 3′ splice site mutations that inhibit the 5′ splice site reaction (106,109), type 2 binding may be inhibited. Binding of unidentified factor(s) at the 5′ splice site, detected by a different nuclease protection assay, is inhibited by such mutations (137). Site-specific hydrolysis of U2 snRNA, which inhibits splicing and binding of U2 snRNP at the branchpoint sequence, inhibits U1 snRNP type 2 binding, but not type 1 binding (113). These observations are consistent with the suggestion that type 2, rather than type 1, binding is the productive interaction. Removal of the 5′ end of U1 snRNA inhibits type 1 binding; surprisingly, type 2 binding still appears to take place under these conditions (113). It seems unlikely, however, that this binding does not require the 5′ terminus of U1. The protection may be due to residual intact U1 snRNA.

The lack of antisera specific for the U4/U6 snRNP has precluded experiments to address whether this snRNP binds pre-mRNA in a sequence-specific manner. As discussed previously, this particle appears to be necessary for the 5′ splice site reaction, and is present in splicing complexes (107,108,111). Site-specific cleavage of the U4 and U6 snRNA moieties, which inhibits splicing, does not appear to change the pre-mRNA protection patterns attributed to U1, U2 and U5 snRNPs (113).

Apparent protection of pre-mRNA sequences by unidentified components was also detected by resistance of specific intron fragments to digestion by ribonuclease A (137). In this case, as before, snRNP binding would only be detected if the relevant snRNA regions are sufficiently resistant to RNase A. ATP-dependent protection surrounding the branchpoint sequence of the first intron of human β-globin was detected. This protection was dependent on an intact 3′ splice site, but not on a 5′ splice site, and may correspond to the U2 snRNP binding mentioned above. It should be noted that the branchpoint region in this intron contains extensive hyphenated dyad symmetry, and hence may be capable of forming a hairpin. This structure could form in an ATP-dependent manner in response to long-range interactions, and would confer RNase A resistance even in the absence of factors bound at the protected site.

It is worthwhile to contrast the above RNase T1 and RNase A protection experiments with the 'footprint' experiments used to detect binding of factors to DNA (243,244). In the latter, chemical or enzymatic probes are used in such a way that each end-labeled target molecule is cleaved once, on average. The protected region is identified as a gap in an electrophoretic ladder of the resulting nested fragments. Detection requires stable binding to all, or nearly all, target molecules. In contrast, the above experiments require extensive enzymatic digestion such that only the protected region survives the treatment. Hence, binding to even a small subpopulation of target molecules can be detected. However, only factors that can stably bind to their cognate regions irrespective of whether adjacent regions are removed can be detected. For example, factors that bind to a specific RNA structure that is sensitive to long range secondary and tertiary interactions will dissociate once most of the RNA is digested.

Since higher order RNA and RNP structures are very likely to be important in pre-mRNA splicing, it is desirable to employ methods to identify binding sites for relevant factors, in a way that overcomes the above problems. Once the components of the splicing apparatus are purified, this type of approach should be very useful to further characterize the factor – pre-mRNA interactions that are necessary for splicing.

Factor binding at or near the branch of the lariat intermediates and products has been detected by measuring protection from enzymatic debranching (110,137). Factor binding has also been probed by RNase H protection experiments employing oligonucleotides complementary to selected regions of the pre-mRNA (137,245). In contrast to the above ribonuclease protection experiments, this method does not result in snRNA degradation, unless the oligonucleotides cross-hybridize with one or more snRNAs. Furthermore, this method does not result in indiscriminate degradation of all unprotected RNA regions, since RNase H cleaves RNA in a site-specific manner. However, this method cannot distinguish between factor binding and inaccessibility due to higher order structure. Finally, whereas the T1 ribonuclease/anti-snRNP antibody protection experiments are specific for snRNP binding, the RNase A and RNase H protection methods are more general, but they do not provide any clues to the nature of the putative binding factors. In any event, the RNase H protection studies detected ATP-dependent, 3′ splice site-dependent binding of a factor at the 5′ splice site, prior to the 5′ splice site reaction (137). This binding may correspond to the type 2 binding of U1 snRNP described above. In addition, early protection of the branchpoint region was also detected, and this protection persisted after lariat formation. In *S.cerevisiae* this method detected ATP-dependent binding of factors at the 5′ splice site and UACUAAC box; the latter protection required an intact 5′ splice site (245).

Filter-binding assays have been used to identify components in HeLa nuclear extracts that bind specifically to 5′ and 3′ splice sites, and to the branchpoint sequence (233,234). DEAE – Sepharose fractions enriched in snRNPs bound RNAs containing 5′ or 3′ splice sites, whereas RNAs containing a consensus branchpoint sequence were bound more weakly. ATP had no effect on these binding interactions. When U1 snRNP was purified more extensively, it only exhibited 5′ splice site binding by this assay. The DEAE flowthrough fraction, which was devoid of snRNAs, contained weak 5′ splice site binding activity, and strong 3′ splice site and branchpoint binding activities. The latter two binding reactions were stimulated by ATP and were resistant to treatment with micrococcal nuclease. Finally, binding of partially purified U1 snRNP to a 5′ splice site was strongly stimulated by addition of the flowthrough fraction. These results indicate that unidentified proteins can stimulate the binding of at least one snRNP to RNA.

Recently, the ability of individual proteins to bind RNA in a sequence-specific manner was examined in blotting experiments (199,200). Nuclear proteins were fractionated by SDS – PAGE and electroblotted onto a filter that was then probed with radiolabeled RNAs. RNA binding proteins with limited or no sequence specificity were tentatively identified as the U1 snRNP-specific polypeptides 68 kd and A. The snRNP D polypeptide also binds many but not all RNAs. More strikingly, an additional polypeptide that binds in a sequence-specific manner to the 3′ splice site polypyrimidine stretch was identified; this binding was enhanced by the presence of the 3′ splice site AG dinucleotide. Tazi *et al.* found that this intron binding protein had an apparent molecular weight of 100 kd on SDS – PAGE, but appeared to be proteolyzed to a polypeptide of 68 kd (199). Gerke and Steitz identified only a 70 kd polypeptide as the 3′ splice site binding protein (200). Both groups reported that this polypeptide appears to associate with U5 snRNP, although it can dissociate in 15 mM $MgCl_2$. In one study, however, the intron binding protein appears to be associated with U5 snRNP after centrifugation in cesium chloride gradients containing 15 mM $MgCl_2$ (199).

Even after dissociation from U5 snRNP, this polypeptide retains Sm epitopes, as shown by immunoprecipitation and Western analysis with a monoclonal anti-Sm antibody (200). This particular antibody also recognizes epitopes in the B', B, D, E and 68 kd polypeptides. Tazi *et al.* did not detect Sm determinants after dissociation of their 100 kd intron binding protein, but this may be due to their use of a different monoclonal antibody, which appears to recognize a less common epitope present only in the D polypeptide (199).

The above results demonstrate that a single polypeptide has the intrinsic ability to bind 3' splice sites in a sequence-specific manner, in the absence of ATP and any additional proteins or snRNAs. It remains to be shown whether this activity is necessary for splicing, and if so, whether this polypeptide has other relevant activities in addition to its RNA binding properties. Under some conditions, this polypeptide appears to be associated with U5 snRNA. U5 as well as the other snRNPs may have dynamic structures, such that their conformation and/or polypeptide composition could vary as they carry out their functions in the pre-mRNA splicing pathway. The reported sizes of the 3' splice site binding proteins are similar to those of the hnRNP polypeptides of 120 kd and 68 kd, which are RNA-binding proteins (reviewed in ref. 6). Whether these polypeptides are related to each other remains to be determined.

7. Alternative splicing

Alternative utilization of 5' and/or 3' splice sites has been observed in a large number of eukaryotic viral and cellular genes (reviewed in ref. 3). In many of these cases, more than one primary transcript can be generated by transcription initiation at alternative sites, and/or by 3' end processing at alternative polyadenylation sites. In such cases, it is possible that the structure of the alternative primary transcripts dictates the subsequent choice of alternative splice sites. Hence, regulation by *trans*-acting factors could be at the level of transcription initiation, or 3' processing. In a few cases, a single primary transcript can be differentially spliced by alternative utilization of 5' and/or 3' splice sites. The choice of the alternative splice sites in this case may be random, in which case alternatively spliced mRNAs would be generated in a fixed proportion, or it may be regulated in a tissue-specific or temporal manner. In the latter case, the choice of alternative splice sites must be the consequence of the action of differentially expressed *trans*-acting factors. *Trans*-acting factors may also regulate the alternative utilization of splice sites in the case of pre-mRNAs that are different at their 5′ or 3′ termini.

Temporal or tissue-specific expression of alternatively spliced mRNAs may be regulated either at the level of mRNA stability, or at the level of splice site choice. In the former case, all alternatively spliced versions of a pre-mRNA may be generated constitutively, but some may be rapidly degraded by specific nucleases whose expression or activity is subject to regulation. In the latter case, there may be positive or negative regulators that modulate the affinity of a general splicing factor for the alternative splice sites, or there may be specialized splicing factors that recognize the relevant splice sites directly. Another possibility is that these putative factors organize the pre-mRNA higher order structure in such a way as to favor the utilization of a particular set of splice sites by the general splicing machinery.

Germ-line restricted expression of the *Drosophila* P-element transposase is an example of alternative splicing regulated at the level of splice site choice (246). One of the introns in the transposase pre-mRNA is spliced in germ-line but not in somatic tissue. Lack of somatic expression is not due to instability of the fully spliced transposase mRNA, since somatic transposition can be forced by expressing an *in vitro*-generated gene in which the relevant intron has been precisely excised. Therefore, regulation of P-element transposase expression depends on whether an intron can be recognised as such in different tissues.

Expression of many other alternatively spliced genes involves utilization of alternative 5′ or 3′ splice sites in a variety of combinations; specific examples are cited in a recent review (3). It should be possible to identify putative *trans*-acting factors responsible for regulated differential splicing, by biochemical complementation using splicing extracts derived from tissues or cell lines in which the alternatively spliced forms are expressed.

8. *Trans* splicing

The usual substrate for RNA splicing is a continuous RNA molecule containing at least two exons and one intron. Under special circumstances, RNA splicing can result in the joining of exons that are part of two separate RNA molecules. This phenomenon is known as intermolecular exon-ligation or *trans* splicing. In all known splicing mechanisms, a cleaved 5′ exon is generated as an intermediate, prior to the final ligation of 5′ and 3′ exons (see Section 2). In pre-tRNA splicing, a cleaved 3′ exon is also generated, whereas in the remaining mechanisms, 3′ splice site cleavage and exon ligation are or appear to be concerted reactions. Therefore, all splicing reactions involve the joining of exons that are part of separate RNA intermediates. Each reaction mechanism ensures that under usual circumstances only exons that are part of the same continuous RNA precursor are joined. To accomplish this, the cleaved exon or exons are held together by non-covalent interactions.

The cleaved 5' and 3' exons generated during pre-tRNA splicing are held together by secondary and tertiary interactions that parallel those of the folded mature tRNA (see Section 2.1). These interactions are strong and specific, since purified, cleaved tRNA exons can base-pair and serve as substrates for the pre-tRNA splicing ligase (12). Therefore, at least the second step of the pre-tRNA splicing reaction can take place as an intermolecular reaction. It is likely that 5' and 3' splice site cleavage would also take place on a pre-tRNA substrate that is cleaved within the intron. Under natural circumstances, however, the base-paired exons are not pulled apart during the cleavage and ligation reactions.

The self-splicing group I introns contain a 5' exon binding site, known as the internal guide sequence, which interacts with the 5' exon by base-pairing (see Section 2.2). The base-pairing interaction positions the 5' exon within the intron active site, where the transesterification reactions take place. This interaction is important for both transesterification steps of the splicing reaction (43). The nature of this base-pairing interaction allows separate RNA molecules that are complementary to the internal guide sequence to function as 5' exons in reactions catalyzed by the intron. Such intermolecular reactions can occur at the level of 5' splice site cleavage and guanosine addition to the 5' end of the intron (247), as well as at the level of 3' splice site cleavage and exon ligation (43). These reactions correspond, respectively, to the first and second transesterifications of the usual group I self-splicing reaction.

A similar 5' exon binding site within a group II intron has been described, although it has not yet been localized to a particular site or sequence (121). The interactions between this unidentified site and the 5' exon are sufficiently strong and specific to allow the group II intron RNA to catalyze intermolecular reactions involving a 5' exon contained in a separate RNA molecule. Both 5' splice site cleavage, as well as 3' splice site cleavage with concomitant exon ligation, were observed as independent reactions. In contrast to the normal group II self-splicing reaction, the *trans*-5' cleavage reaction did not proceed by transesterification but rather by hydrolysis of the 5' splice site phosphodiester bond, and hence no lariat intermediate was formed (121).

Intermolecular splicing reactions have also been obtained with nuclear pre-mRNA *in vitro* splicing systems (248,249). In this case, the substrate consists of two RNAs that are held together by a complementary region. One RNA molecule consists of a 5' exon and a fragment of the adjacent intron extending beyond the 5' splice site; the 3' end segment of this RNA is complementary to the 5' terminus of the second RNA molecule. This latter RNA also contains a branchpoint sequence, 3' splice site and adjacent 3' exon. The hybridized RNAs resemble a natural pre-mRNA substrate, except for the presence of a hairpin and a break in the polynucleotide chain within the intron. This substrate undergoes a complete splicing reaction; the cut within the intron is not repaired, and hence the intron is excised as a branched, linear RNA. Base-pairing between the two RNA fragments appears to be necessary for *trans*-splicing under these conditions. However, the base-pairing capability was deliberately engineered into the RNA substrates, and does not reflect known general features of usual pre-mRNA substrates. Inefficient *trans*-splicing was obtained with RNAs that lacked extensive complementarity (249), but the possibility that intermolecular splicing of these substrates depends on limited base-pairing has not been ruled out. In the standard pre-mRNA splicing reaction, the cleaved 5' exon is part of the same RNP complex as the intron – 3' exon intermediate, and hence there must be stable interactions that hold the two RNA splicing intermediates together (82,83). The nature of these interactions is unknown, but they may consist of RNA – RNA, RNA – protein, and/or protein – protein interactions between different regions of the pre-mRNA, various snRNAs, and protein factors. Whether 5' and 3' exons in separate RNAs can come together solely by these kinds of interactions, to serve as substrates for intermolecular exon joining, remains to be determined.

Natural examples of putative *trans*-splicing reactions have been identified. Trypanosome mRNAs are assembled from two transcripts derived from separate genes (reviewed in refs 250,251). A 35 nucleotide capped leader is found at the 5' end of all mRNAs; it is derived from a longer RNA precursor encoded in a multigene family. The structures of the primary transcripts encoding each mRNA 3' portion are not known, but they may be part of polycistronic transcripts. Analysis of the genes encoding various mRNA 3' segments shows putative 3' splice sites upstream of each coding segment. Likewise, the RNA precursor from which the capped leader is derived contains a putative 5' splice site downstream of the mature leader sequences. Therefore, it appears that mRNAs are generated by splicing of the leader, or mini-exon, to any of various separately encoded 3' exons. The splicing substrate could be generated by transcription of the 3' half of each pre-mRNA using the mini-exon primary transcript or a 5' fragment of it as a primer. Alternatively, the mini-exon and 3' exon-containing primary transcripts are ligated, creating a standard pre-mRNA splicing substrate. Finally, the same primary transcripts may remain separate, and the exons may be joined by *trans*-splicing. Since all of the above models involve splicing, they all predict the existence of a free mini-exon intermediate RNA of 35 nucleotides. Such a species has been identified (252). In addition, the last model predicts the existence of branched linear molecules as intermediates and products of the reaction. The recent identification of a 100 nucleotide 3' terminal fragment of the mini-exon primary transcript upon treatment of high molecular weight RNA with debranching enzyme is consistent with the existence of such branched linear structures (252 – 254). The sites of branch formation have not been identified, but based on the above result

they would have to be located within each of the RNAs containing a 3′
splice site and 3′ exon. The interpretation of these results is complicated
by the observation that similar mini-exon transcript fragments of 100 and
35 nucleotides are generated by spontaneous or artifactual cleavage of
the primary transcript (252). Therefore, the fact that such fragments are
also released by treatment of the high molecular weight RNA with crude
(252,254) or partially purified (253) preparations of debranching enzyme
is not sufficient demonstration of the existence of branched linear RNAs
(252). Direct identification of these structures, and the demonstration that
they constitute intermediates in the generation of mRNA, will be necessary to confirm the *trans*-splicing model.

   If the *trans*-splicing model of trypanosome discontinuous mRNA
synthesis is correct, it will be of interest to determine the nature of the
non-covalent interactions that bring together the two primary transcripts.
Complementary regions in these RNAs have not been found, and hence
these interactions may be mediated by cellular factors. Except for these
discontinuous introns, other introns have not been found in trypanosome
genes analyzed to date, and therefore *trans*-splicing may be the only form
of splicing used by trypanosomes. A trypanosome U2 snRNA homolog
has recently been identified (255); complementarity between a region of
this snRNA and the mini-leader transcript has been pointed out (255).
It will be of interest to determine whether this and other trypanosome
snRNAs are involved in the proposed *trans*-splicing reaction, and how
their involvement relates to the function of metazoan and yeast snRNAs
in the standard pre-mRNA splicing reaction.

   Trans-splicing has also been proposed as a mechanism for the synthesis
of the mRNA for ribosomal protein S12 in the chloroplasts of *Nicotiana
tabacum* and *Marchantia polymorpha* (256–259). The coding sequence
corresponding to the amino terminus of the protein is found as far as 90 kb
away from the 3′ portion of the gene, on the circular chloroplast genome.
Furthermore, the putative 5′ exon is sometimes found in opposite
orientation relative to the rest of the gene. Electron microscopy studies
also indicate that the two gene segments are part of separate transcription
units (259). The adjacent intron sequences suggest a mechanism related
to the splicing of group II introns.

9. Evolution of RNA splicing

Since the discovery of RNA splicing in 1977 (260,261), it has become clear
that the organization of cellular and viral transcription units into exons
and introns is prevalent among higher eukaryotes. Although in some cases
they are less common, introns have been found in the transcription units
of animals, plants, fungi and protista. A few examples of interrupted genes
are known in pre-tRNA (262,264) and pre-rRNA (266) genes of

archaebacteria, and in pre-mRNA genes of a eubacterial phage (40,41). In eukaryotes, introns have been found in genes encoding pre-mRNA, pre-rRNA and pre-tRNA, in nuclei, mitochondria and chloroplasts. The widespread distribution of introns in such diverse genomes indicates that either interrupted genes were present in primordial organisms, prior to the divergence between prokaryotes and eukaryotes, and/or that mechanisms for the transmission of introns and their insertion into new locations existed later during evolution.

Although introns are excised by several different mechanisms (see Section 2), the above generalization applies at least to the group I introns (see Section 2.2), which are clearly related by sequence, structure and mechanism of excision, and are found in pre-rRNA, pre-mRNA and pre-tRNA of plants, fungi, protista and eubacteria, in nuclear, mitochondrial, chloroplast and bacteriophage genomes. In addition, functional similarities among different splicing mechanisms, and/or in the location of different kinds of introns, suggest that the different RNA splicing mechanisms known today may have evolved from a primordial form of RNA splicing.

RNA catalysis is thought to have played a critical role during pre-biotic evolution (reviewed in refs 4,31,263,265). This role may be reflected in such contemporary RNA catalyzed reactions as pre-tRNA processing by the RNA subunit of RNase P (reviewed in refs 30,31), self-splicing of group I and group II introns (see Sections 2.1 and 2.2), self-dimerization and self-cleavage of plant RNA viroids and virusoids (reviewed in ref. 4), and self-cleavage of the transcripts of the newt satellite DNA (267). Other reactions involving essential RNA components, which may or may not have a catalytic role, include translation by ribosomes, which contain rRNA subunits; protein translocation by the signal recognition particle, which contains 7SL RNA (268); sea-urchin histone 3' processing, which requires U7 snRNA (reviewed in ref. 144, Chapter 4); pre-mRNA splicing, which requires several small RNAs (see Section 6.1); polysaccharide branching by the rabbit muscle amylose isomerase, which is associated with a 31 nt RNA (269); initiation of human mitochondrial DNA replication by DNA primase, which is associated with RNAs that are essential for activity (270); catalysis of thiol-disulfide exchange by a ribonucleoprotein from sea urchin eggs (271); synthesis of δ-aminolevulinic acid by an enzyme from *Chlamydomonas* that is associated with essential RNAs (272); formation of *Tetrahymena* telomere ends, which requires an unidentified RNA component (C.W.Greider and E.H.Blackburn, in press).

In a pre-biotic scenario in which RNA catalyzed reactions are prevalent, self-splicing introns may have been present. Despite similarities in their self-splicing mechanisms, present day group I and group II introns have different conserved sequences and structures. It is not known if they evolved from a common ancestor or if they arose independently. It has been shown that a group I intron that self-splices by the standard group

I mechanism can also generate lariat structures by autocatalysis (273). Although their structures indicate that these lariat RNAs are not splicing intermediates, these results show that an RNA with group I structure and sequence can undergo transesterifications involving an internal 2'-hydroxyl group, a reaction that characterizes the self-splicing of the group II introns. Whatever the evolutionary relation between group I and group II introns, both forms of self-splicing involve transesterification reactions. Therefore, intron insertion, the reverse of self-splicing, is thermodynamically permissible (44,263). Similar reactions may have taken place during evolution, and reverse transcription, recombination, and gene conversion events may account for the eventual appearance of introns in DNA, thus ensuring their stable inheritance.

The group I intron of the S.cerevisiae mitochondrial large rRNA can efficiently insert itself at the corresponding site of intronless genes during genetic crosses. Interestingly, this insertion requires the product of an intron-encoded open reading frame (274,275). Although this product is not required for splicing of the L-rRNA intron, it is homologous to maturases. The mechanism of insertion is not known, but it could involve insertion at the RNA level, followed by reverse transcription and gene conversion, or it could involve some form of DNA transposition not involving RNA intermediates. Related, though more permissive, mechanisms may have operated during evolution to spread introns throughout primitive genomes. As long as the introns retained the necessary information for their own excision at the RNA level, they did not impair gene expression. At this stage, some introns may have conferred selective advantages, or they may have propagated as selfish elements without damage to the host (276,277).

Eventually, protein and/or RNA trans-acting factors may have evolved with the capacity to regulate, or in some cases catalyze, the splicing reactions. The existence of such factors would allow the loss of the catalytic domains of self-splicing introns; only the structural information necessary for precise excision would remain. Present day examples of introns that are clearly related to self-splicing introns, judging from their sequence and structure and from the nature of the splicing intermediates, but require trans-acting factors for their excision, can be found among type I and type II introns (see Sections 2.2, 2.3 and 2.5). Nuclear pre-mRNA introns may also be related to self-splicing type II introns, considering the similarity in the structures of the lariat intermediates generated during both splicing reactions. In this case, the two types of introns appear to have diverged extensively, perhaps reflecting the dependence of nuclear pre-mRNA splicing on multiple cellular factors.

Present day nuclear pre-tRNA introns, which are found at a conserved position in the anticodon loop, lack conserved sequences and structures, and are as small as eight nucleotides (see Section 2.1). The information necessary for their precise removal is found primarily in highly conserved features of the pre-tRNA exons. If this class of introns also evolved from primordial self-splicing introns, the protein enzymes that eventually took over the catalytic role may have evolved to recognize conserved features common to all pre-tRNA exons. This would in turn allow the introns to diverge extensively in size and sequence. The existence of long group I and group II introns in the anticodon loops of some chloroplast pre-tRNAs is consistent with the suggestion that a common ancestor of all interrupted tRNA genes may have contained a self-splicing intron.

In summary, we can envisage plausible mechanisms for the appearance and spread of introns in primitive genomes, and even prior to the appearance of the first life forms. Some contemporary organisms, such as the prokaryotes, have few or no introns; it is thought that during their evolutionary history they have streamlined their genomes to allow more efficient growth (278,279). Intron loss can be observed at the DNA level in group I and group II introns of mitochondrial genomes (280). Its occurrence during evolution can be inferred from phylogenetic comparisons of the intron locations in related genes and gene families (reviewed in refs 281–283). These phylogenetic comparisons suggest that primitive pre-mRNA genes contained many more introns than they do today. Plausible mechanisms for intron loss, involving reverse transcription of spliced transcripts followed by gene conversion, have been proposed (reviewed in ref. 263), and are supported by the structures of processed pseudogenes. With a pre-mRNA splicing machinery already in place, intron insertions may also have occurred during relatively recent evolutionary history (284), although mechanisms for the insertion of nuclear pre-mRNA introns into new locations are difficult to envisage.

The biological and evolutionary significance of introns is unclear. The discovery of RNA catalysis has led to the suggestion that introns represent evolutionary relics of pre-biotic mechanisms for self-propagating RNA (29,30,263,265,277). Introns have not just propagated selfishly, since their hosts are likely to have derived selective advantage from having introns and an RNA splicing machinery. Introns and RNA splicing could have participated in many ways to create new genes, thus conferring selective advantage. Assortment at the DNA and/or RNA level of exon segments encoding discrete functional or structural polypeptide domains has been proposed as a mechanism for generating diversity during evolution (265,281,282). In agreement with this view, many introns separate exons encoding functional and/or structural protein domains or modules (reviewed in refs 282,283). Furthermore, evolutionarily related protein domains are sometimes found encoded in exons that are shared by otherwise unrelated genes (reviewed in ref. 265).

At least some contemporary introns do not appear to affect expression of the genes containing them (167,168,280,285). Retention of such introns may reflect the lack of selective advantage resulting from their removal, or it may reflect inefficiency in the mechanisms of intron loss. In at least some cases, introns have been retained because they serve important functions in the expression of certain genes. A few known examples of such functions are described below.

Many examples are now known of eukaryotic pre-mRNAs that are alternatively spliced, resulting in the expression of related proteins (reviewed in ref. 3; Section 7). Expression of the full coding potential of these genes requires alternative splicing, thus creating selective pressure for retaining at least some introns and a pre-mRNA splicing machinery. In some cases, alternative splicing may be regulated in a temporal or tissue-specific manner (see Section 7). Expression of related proteins encoded in a single transcription unit by alternative splicing is a way to generate diversity. This strategy may have been even more common during early evolution than it is today. An alternative way to generate diversity is gene duplication, resulting in the expression of related proteins encoded by different members of a gene family. The former arrangement requires less DNA, and is commonly found in compact viral genomes. Gene duplication may allow greater flexibility in generating novel functions, because all regions of the duplicated gene can diverge without impairing the ancestral function. Nevertheless, both evolutionary strategies are common, and sometimes are used in combination, as judged from the organization of many genes.

In addition to the regulation of splice site choice, there are other ways in which splicing can be used to regulate gene expression. Splicing of several pre-mRNAs encoding ribosomal proteins appears to be inhibited by the corresponding ribosomal protein (286; reviewed in ref. 287). It has been proposed that this feedback mechanism is used for autogenous control of ribosomal protein expression. A few fungal mitochondrial group I and group II introns contain open reading frames whose protein products, known as maturases, are involved in the excision of the introns encoding them (see Sections 2.2, 2.3 and 2.5). This situation may be used for feedback regulation of the entire transcription unit (37).

Other steps in RNA biogenesis have co-evolved with splicing. As a result, in some cases, RNA processing reactions other than splicing may require intron-containing RNAs as obligatory substrates. For example, the anticodon pseudouridine modification of *S.cerevisiae* pre-tRNA^{Tyr} requires the presence of the intron (288). cDNAs derived from the mRNAs of some eukaryotic intron-containing genes are not expressed when transfected on suitable expression vectors unless they carry homologous or heterologous introns (164 – 166). In these particular cases, RNA stability or efficient nuclear – cytoplasmic transport could be dependent on the presence of introns. However, this is not a general phenomenon, since examples of genes whose introns are removed without affecting expression are known (167,168,285).

Finally, some introns contain important information at the DNA level. For example the immunoglobulin heavy chain enhancer is a *cis*-acting transcriptional regulatory element located in the intron upstream of the constant region segment. This position allows the enhancer to direct transcription from any of the variable region promoters brought into the vicinity of the constant region by DNA rearrangement (289–291). In addition to the group I and group II introns that encode maturases, a variety of other introns encode open reading frames. The functions of these intron products are unknown (41; reviewed in ref. 39), except in one case, in which the intron-encoded protein is involved in transposition of that intron (274,275). An example of nested genes transcribed in opposite directions has been documented, in which a gene is entirely contained within an intron of a larger gene (292).

10. Concluding remarks

In the ten years following the discovery of introns in viral pre-mRNAs, substantial progress has been made in attempting to understand the mechanisms, function and evolutionary significance of RNA splicing. Introns have been shown to be widespread in a variety of genomes, and four distinct types of splicing mechanisms have been partly elucidated. Sequence comparisons and mutational analyses have helped delineate pre-RNA structural features that are important for the splicing mechanisms and for the specificity of the reactions. Cell-free systems have been developed for each of the four types of RNA splicing reactions known to date. These systems have made it possible to characterize the biochemical requirements of the reactions, as well as the structures of splicing intermediates.

The splicing mechanisms of group I and group II introns involve RNA catalysis, and several members of these two classes of introns are capable of self-splicing, although proteins can influence the rate of these reactions *in vivo*. In contrast, nuclear pre-tRNA splicing is catalyzed by protein enzymes, whereas nuclear pre-mRNA splicing requires multiple RNA and protein factors. However, all splicing mechanisms may have evolved from an ancestral self-splicing mechanism. Nuclear pre-mRNA introns, which constitute by far the largest class of introns in present day eukaryotes, appear to be related to the type II introns, which so far have been found only in fungal and plant organelles. There is limited similarity at the level of sequence, and substantial similarity in the structures of the respective reaction intermediates. There may be further similarities in the use of RNA–RNA interactions and quite possibly RNA catalysis, but this remains to be determined.

Elucidation of the detailed pathway of nuclear pre-mRNA splicing will require the purification and characterization of the factors that catalyze the reaction. However, the complexity of the nuclear pre-mRNA splicing apparatus may rival that of the translational machinery, and hence, the isolation of all the components of the splicing apparatus may be an arduous task. Further rapid progress in this area is anticipated in *S.cerevisiae* systems, in which the combination of biochemical and genetic approaches has proved to be especially powerful.

The study of two essential aspects of the mechanism of nuclear pre-mRNA splicing should be facilitated by the availability of purified components. The first aspect includes the nature and mechanism of action of the ribozymes and/or enzymes that catalyze the reaction. The second aspect of the splicing mechanism is the molecular basis of the specificity of splice site selection. Our present understanding of the reaction is insufficient to account for its high degree of specificity. This latter aspect also relates to the utilization of alternative splice sites in a wide variety of complex transcription units. Regulation of splice site selection in response to spatial, temporal, or physiological state clues may be a common strategy for the regulation of gene expression. Future work may uncover these regulatory factors and lead to an understanding of how they interact with the pre-mRNA substrates and with the general splicing apparatus.

Finally, the identification and characterization of the components of the nuclear pre-mRNA splicing apparatus, as well as a more detailed understanding of all splicing mechanisms, may uncover evolutionary relations among different introns and catalytic entities, thus leading to a greater understanding of the origins and functions of RNA splicing.

11. Acknowledgements

We thank our colleagues in splicing for sharing information and ideas. We are especially grateful to David Frendeway and Rich Roberts for valuable discussions and for helpful comments on the manuscript.

12. References

1. Padgett,R.A., Grabowski,P.J., Konarska,M.M., Seiler,S. and Sharp,P.A. (1986) Splicing of messenger RNA precursors. *Annu. Rev. Biochem.*, **55**, 1119.
2. Green,M.R. (1986) Pre-mRNA splicing. *Annu. Rev. Genet.*, **20**, 671.
3. Leff,S.E., Rosenfeld,M.G. and Evans,R.M. (1986) Complex transcriptional units: diversity in gene expression by alternative RNA processing. *Annu. Rev. Biochem.*, **55**, 1091.
4. Cech,T.R. and Bass,B.L. (1986) Biological catalysis by RNA. *Annu. Rev. Biochem.*, **55**, 599.
5. Brunel,C., Sri-Widada,J. and Jeanteur,P. (1985) snRNP's and scRNP's in eukaryotic cells. *Prog. Mol. Subcell. Biol.*, **9**, 1.
6. Dreyfuss,G. (1986) Structure and function of nuclear and cytoplasmic ribonucleoprotein particles. *Annu. Rev. Cell Biol.*, **2**, 459.
7. Abelson,J. (1979) RNA processing and the intervening sequence problem. *Annu. Rev. Biochem.*, **48**, 1035.
8. Sprinzl,M., Hartmann,T., Meissner,F., Moll,J. and Vorderwulbecke,T. (1987)

Compilation of tRNA sequences and sequences of tRNA genes. *Nucleic Acids Res.,* **15** (Suppl.), r53.

9. Swerdlow,H. and Guthrie,C. (1984) Structure of intron-containing tRNA precursors. *J. Biol. Chem.,* **259**, 5197.

10. Lee,M.-C. and Knapp,G. (1985) Transfer RNA splicing in *Saccharomyces cerevisiae. J. Biol. Chem.,* **260**, 3108.

11. Greer,C.L., Söll,D. and Willis,I. (1987) Substrate recognition and identification of splice sites by the tRNA-splicing endonuclease and ligase from *Saccharomyces cerevisiae. Mol. Cell. Biol.,* **7**, 76.

12. Peebles,C.L., Ogden,R.C., Knapp,G. and Abelson,J. (1979) Splicing of yeast tRNA precursors: a two-stage reaction. *Cell,* **18**, 27.

13. Peebles,C.L., Gegenheimer,P. and Abelson,J. (1983) Precise excision of intervening sequences from precursor tRNAs by a membrane-associated yeast endonuclease. *Cell,* **32**, 525.

14. Greer,C.L., Peebles,C.L., Gegenheimer,P. and Abelson,J. (1983) Mechanism of action of a yeast RNA ligase in tRNA splicing. *Cell,* **32**, 537.

15. Phizicky,E.M., Schwartz,R.C. and Abelson,J. (1986) *Saccharomyces cerevisiae* tRNA ligase. Purification of the protein and isolation of the structural gene. *J. Biol. Chem.,* **261**, 2978.

16. Konarska,M., Filipowicz,W., Domdey,H. and Gross,H.J. (1981) Formation of a 2'-phosphomonoester, 3',5'-phosphodiester linkage by a novel RNA ligase in wheat germ. *Nature,* **293**, 112.

17. Konarska,M., Filipowicz,W. and Gross,H.J. (1982) RNA ligation via 2'-phosphomonoester, 3',5'-phosphodiester linkage: requirement of 2',3'-cyclic phosphate termini and involvement of a 5'-hydroxyl polynucleotide kinase. *Proc. Natl. Acad. Sci. USA,* **79**, 1474.

18. Cranston,J.W., Silber,R., Malathi,V.G. and Hurwitz,J. (1974) Studies on ribonucleic acid ligase. *J. Biol. Chem.,* **249**, 7447.

19. Ohtsuka,E., Nishikawa,S., Sugiura,M. and Ikehara,M. (1976) Joining of ribo-oligonucleotides with T4 RNA ligase and identification of the oligonucleotide-adenylate intermediate. *Nucleic Acids Res.,* **3**, 1613.

20. Sugino,A., Snopek,T.J. and Cozzarelli,N.R. (1977) Bacteriophage T4 RNA ligase: reaction intermediates and interaction of substrates. *J. Biol. Chem.,* **252**, 1732.

21. Nishikura,K. and DeRobertis,E.M. (1981) RNA processing in microinjected *Xenopus* oocytes. Sequential addition of base modifications in the spliced transfer RNA. *J. Mol. Biol.,* **145**, 405.

22. Filipowicz,W. and Shatkin,A.J. (1983) Origin of splice junction phosphate in tRNAs processed by HeLa cell extract. *Cell,* **32**, 547.

23. Laski,F.A., Fire,A.Z., RajBhandary,U.L. and Sharp,P.A. (1983) Characterization of tRNA precursor splicing in mammalian extracts. *J. Biol. Chem.,* **258**, 11974.

24. Van Tol,H., Stange,N., Gross,H.J. and Beier,H. (1987) A human and a plant intron-containing tRNA^Tyr gene are both transcribed in a HeLa cell extract but spliced along different pathways. *EMBO J.,* **6**, 35.

25. Filipowicz,W., Konarska,M., Gross,H.J. and Shatkin,A.J. (1983) RNA 3' terminal phosphate cyclase activity and RNA ligation in HeLa cell extract. *Nucleic Acids Res.,* **11**, 1405.

26. Perkins,K.K., Furneaux,H. and Hurwitz,J. (1985) Isolation and characterization of an RNA ligase from HeLa cells. *Proc. Natl. Acad. Sci. USA,* **82**, 684.

27. Reinberg,D., Arenas,J. and Hurwitz,J. (1985) The enzymatic conversion of 3' phosphate terminated RNA chains to 2',3'-cyclic phosphate derivatives. *J. Biol. Chem.,* **260**, 6088.

28. Cech,T.R., Zaug,A.J. and Grabowski,P.J. (1981) *In vitro* splicing of the ribosomal RNA precursor of *Tetrahymena*: involvement of a guanosine nucleotide in the excision of the intervening sequence. *Cell,* **27**, 487.

29. Kruger,K., Grabowski,P.J., Zaug,A.J., Sands,J., Gottschling,D.E. and Cech,T.R. (1982) Self-splicing RNA: autoexcision and autocyclization of the ribosomal RNA intervening sequence of *Tetrahymena. Cell,* **31**, 147.

30. Altman,S. (1984) Aspects of biochemical catalysis. *Cell,* **36**, 237.

31. Waugh,D.S. and Pace,N.R. (1986) Catalysis by RNA. *Bioessays,* **4**, 56.

32. Nomiyama,H., Sakaki,Y. and Takagi,Y. (1981) Nucleotide sequence of a ribosomal

RNA gene intron from slime mold *Physarum polycephalum*. *Proc. Natl. Acad. Sci. USA,* **78**, 1376.

33. Michel,F., Jacquier,A. and Dujon,B. (1982) Comparison of fungal mitochondrial introns reveals extensive homologies in RNA secondary structure. *Biochimie,* **64**, 867.

34. Michel,F. and Dujon,B. (1983) Conservation of RNA secondary structure in two intron families including mitochondrial-, chloroplast-, and nuclear-encoded members. *EMBO J.,* **2**, 33.

35. Davies,R.W., Waring,R.B., Ray,J.A., Brown,T.A. and Scazzochio,C. (1982) Making ends meet: a model for RNA splicing in fungal mitochondria. *Nature,* **300**, 719.

36. Tabak,H.F. and Grivell,L.A. (1986) RNA catalysis in the excision of yeast mitochondrial introns. *Trends Genet.,* **2**, 51.

37. Lazowska,J., Jacq.C. and Slonimski,P.P. (1980) Sequence of introns and flanking exons in wild-type and box3 mutants of cytochrome b reveals an interlaced splicing protein coded by an intron. *Cell,* **22**, 333.

38. Bonnard,G., Michel,F., Weil,J.H. and Steinmetz,A. (1984) Nucleotide sequence of the split tRNA^{Leu} (UAA) gene from *Vicia faba* chloroplasts: evidence for structural homologies of the chloroplast tRNA^{Leu} intron with the intron from the autospliceable *Tetrahymena* ribosomal RNA precursor. *Mol. Gen. Genet.,* **194**, 330.

39. Shinozaki,K., Deno,H., Sugita,M., Kuramitsu,S. and Sugiura,M. (1986) Intron in the gene for the ribosomal protein S16 of the tobacco chloroplast and its conserved boundary sequences. *Mol. Gen. Genet.,* **202**, 1.

40. Chu,F.K., Maley,G.F., Maley,F. and Belfort,M. (1984) An intervening sequence in the thymidylate synthase gene of bacteriophage T4. *Proc. Natl. Acad. Sci. USA,* **81**, 3049.

41. Gott,J.M., Shub,D.A. and Belfort,M. (1986) Multiple self-splicing introns in bacteriophage T4; evidence from autocatalytic GTP labeling of the RNA *in vitro. Cell,* **47**, 81.

42. Cech,T.R. (1986) The generality of self-splicing RNA: relationship to nuclear mRNA splicing. *Cell,* **44**, 207.

43. Been,M.D. and Cech,T.R. (1986) One binding site determines sequence specificity of *Tetrahymena* pre-rRNA self-splicing, trans-splicing, and RNA enzyme activity. *Cell,* **47**, 207.

44. Zaug,A.J., Grabowski,P.J. and Cech,T.R. (1983) Autocatalytic cyclization of an excised intervening sequence RNA is a cleavage-ligation reaction. *Nature,* **301**, 578.

45. Garriga,G. and Lambowitz,A.M. (1984) RNA splicing in *Neurospora* mitochondria: self-splicing of a mitochondrial intron *in vitro. Cell,* **38**, 631.

46. Garriga,G. and Lambowitz,A.M. (1986) Protein-dependent splicing of a group I intron in ribonucleoprotein particles and soluble fractions. *Cell,* **46**, 669.

47. Van der Horst,G. and Tabak,H.F. (1985) Self-splicing of yeast mitochondrial ribosomal and messenger RNA precursors. *Cell,* **40**, 759.

48. Wollenzien,P.L., Cantor,C.R., Grant,D.M. and Lambowitz,A.M. (1983) RNA splicing in *Neurospora* mitochondria: structure of the unspliced 35S precursor ribosomal RNA detected by psoralen cross-linking. *Cell,* **32**, 397.

49. Guerrier-Takada,C., Gardiner,K., Marsh,T., Pace,N. and Altman,S. (1983) The RNA moiety of ribonuclease P is the catalytic subunit of the enzyme. *Cell,* **35**, 849.

50. Keller,M. and Michel,F. (1985) The introns of the *Euglena gracilis* chloroplast gene which codes for the 32-kDa protein of photosystem II. *FEBS Lett.,* **179**, 69.

51. Cech,T.R. (1983) RNA splicing: three themes with variations. *Cell,* **34**, 713.

52. Schmelzer,C. and Schweyen,R.J. (1986) Self-splicing of group II introns *in vitro*: mapping of the branch point and mutational inhibition of lariat formation. *Cell,* **46**, 557.

53. Peebles,C.L., Perlman,P.S., Mecklenburg,K.L., Petrillo,M.L., Tabor,J.H., Jarrell,K.A. and Cheng,H.-L. (1986) A self-splicing RNA excises an intron lariat. *Cell,* **44**, 213.

54. Van der Veen,R., Arnberg,A.C., Van der Horst,G., Bonen,L., Tabak,H.F. and Grivell,L.A. (1986) Excised group II introns in yeast mitochondria are lariats and can be formed by self-splicing *in vitro. Cell,* **44**, 225.

55. Schmelzer,C., Schmidt,C., May,K. and Schweyen,R.J. (1983) Determination of functional domains in intron bl1 by study of yeast mitochondrial mutations and a nuclear suppressor. *EMBO J.,* **2**, 2047.

56. Carignani,G., Groudinsky,O., Frezza,D., Schiavon,E., Bergantino,E. and

Slonimski,P.P. (1983) An mRNA maturase is encoded by the first intron of the mitochondrial gene for the subunit I of cytochrome oxidase in *S.cerevisiae. Cell,* **35**, 733.

57. Ghosh,P.K., Reddy,V.B., Swinscoe,J., Lebovitz,P. and Weissman,S.M. (1978) Heterogeneity and 5'-terminal structures of the late RNAs of simian virus 40. *J. Mol. Biol.,* **126**, 813.

58. Scott,M.P., Weiner,A.J., Hazelrigg,T.I., Polisky,B.A., Pirrota,V., Scalenghe,F. and Kaufman,T.C. (1983) The molecular organization of the *Antennapedia* locus of *Drosophila. Cell,* **35**, 763.

59. Schnuewly,S., Kuroiwa,A., Baumgartner,P. and Gehring,W.J. (1986) Structural organization and sequence of the homeotic gene *Antennapedia* of *Drosophila melanogaster. EMBO J.,* **5**, 733.

60. Breathnach,R., Benoist,C., O'Hare,K., Gannon,F. and Chambon,P. (1978) Ovalbumin gene: evidence for a leader sequence in mRNA and DNA sequences at the exon-intron boundaries. *Proc. Natl. Acad. Sci. USA,* **75**, 4853.

61. Mount,S.M. (1982) A catalogue of splice junction sequences. *Nucleic Acids Res.,* **10**, 459.

62. Ruskin,B., Krainer,A.R., Maniatis,T. and Green,M.R. (1984) Excision of an intact intron as a novel lariat structure during pre-mRNA splicing *in vitro. Cell,* **38**, 317.

63. Zeitlin,S. and Efstratiadis,A. (1984) *In vitro* splicing products of the rabbit β-globin pre-mRNA. *Cell,* **39**, 589.

64. Keller,E.B. and Noon,W.A. (1984) Intron splicing: a conserved internal signal in introns of animal pre-mRNAs. *Proc. Natl. Acad. Sci. USA,* **81**, 7417.

65. Brown,J.W.S. (1986) A catalogue of splice junction and putative branch point sequences from plant introns. *Nucleic Acids Res.,* **14**, 9549.

66. Miller,A.M. (1984) The yeast MATa1 gene contains two introns. *EMBO J.,* **3**, 1061.

67. Langford,C.J., Klinz,F.J., Donath,C. and Gallwitz,D. (1984) Point mutations identify the conserved intron-contained TACTAAC box as an essential splicing signal sequence in yeast. *Cell.,* **36**, 645.

68. Leer,R.J., Van Raamsdonk-Duin,M.M.C., Hagendoorn,M.J.M., Mager,W.H. and Planta,R.J. (1984) Structural comparison of yeast ribosomal protein genes. *Nucleic Acids Res.,* **12**, 6685.

69. Teem,J.L., Abovich,N., Kaufer,N.F., Schwindinger,W.F., Warner,J.R., Levy,A., Woolford,J., Leer,R.J., van Raamsdonk-Duin,M.M.C., Mager,W.H., Planta,R.J., Schultz,L., Friesen,J.D., Fried,H. and Rosbash,M. (1984) A comparison of yeast ribosomal protein gene DNA sequences. *Nucleic Acids Res.,* **12**, 8295.

70. Russell,P. and Nurse,P. (1986) *Schizosaccharomyces pombe* and *Saccharomyces cerevisiae:* a look at yeasts divided. *Cell,* **45**, 781.

71. Treisman,R., Orkin,S.H. and Maniatis,T. (1983) Structural and functional defects in β-thalassemia. In *Globin Gene Expression and Hematopoietic Differentiation.* Stamatoyannopoulos,G. and Nienhuis,A. (eds), Alan R. Liss, Inc., New York, p. 99.

72. Maniatis,T. and Reed,R. (1987) The role of small nuclear ribonucleoprotein particles in pre-mRNA splicing. *Nature,* **325**, 673.

73. Beggs,J.D., Van den Berg,J., Van Ooyen,A. and Weissman,C. (1980) Abnormal expression of chromosomal rabbit β-globin gene in *Saccharomyces cerevisiae. Nature,* **283**, 835.

74. Watts,F., Castle,C. and Beggs,J. (1983) Aberrant splicing of *Drosophila* alcohol dehydrogenase transcripts in *Saccharomyces cerevisiae. EMBO J.,* **2**, 2085.

75. Langford,C., Nellen,W., Niesing,J. and Gallwitz,D. (1983) Yeast is unable to excise foreign intervening sequences from hybrid gene transcripts. *Proc. Natl. Acad. Sci. USA,* **80**, 1496.

76. Kaufer,N.F., Simanis,V. and Nurse,P. (1985) Fission yeast *Schizosaccharomyces pombe* correctly excises a mammalian RNA transcript intervening sequence. *Nature,* **318**, 78.

77. Ruskin,B., Pikielny,C.W., Rosbash,M. and Green,M.R. (1985) Alternative branchpoints are selected during splicing of a yeast pre-mRNA in mammalian and yeast extracts. *Proc. Natl. Acad. Sci. USA,* **83**, 2022.

78. Brown,J.W.S., Feix,G. and Frendeway,D. (1986) Accurate *in vitro* splicing of two pre-mRNA plant introns in a HeLa cell nuclear extract. *EMBO J.,* **5**, 2749.

79. Barta,A., Sommergruber,K., Thompson,D., Hartmuth,K., Matzke,M.A. and

Matzke,A.J.M. (1986) The expression of a nopaline synthase-human growth hormone chimaeric gene in transformed tobacco and sunflower callus tissue. *Plant Mol. Biol.*, **6**, 347.

80. Hartmuth,K. and Barta,A. (1986) *In vitro* processing of a plant pre-mRNA in a HeLa cell nuclear extract. *Nucleic Acids Res.*, **14**, 7513.

81. Brody,E. and Abelson,J. (1985) The 'spliceosome': yeast pre-messenger RNA associates with a 40S complex in a splicing-dependent reaction. *Science*, **228**, 963.

82. Grabowski,P.J., Seiler,S.R. and Sharp,P.A. (1985) A multi-component complex is involved in the splicing of messenger RNA precursors. *Cell*, **42**, 345.

83. Frendeway,D. and Keller,W. (1985) Stepwise assembly of a pre-mRNA splicing complex requires U-snRNPs and specific intron sequences. *Cell*, **42**, 355.

84. Hernandez,N. and Keller,W. (1983) Splicing of *in vitro* synthesized messenger RNA precursors in HeLa cell extracts. *Cell*, **35**, 89.

85. Hardy,S.F., Grabowski,P.J., Padgett,R.A. and Sharp,P.A. (1984) Cofactor requirements for splicing of purified messenger RNA precursors. *Nature*, **308**, 375.

86. Krainer,A.R., Maniatis,T., Ruskin,B. and Green,M.R. (1984) Normal and mutant human β-globin pre-mRNAs are faithfully and efficiently spliced *in vitro. Cell*, **36**, 993.

87. Lin,R.-J., Newman,A.J., Cheng,S.-C. and Abelson,J. (1985) Yeast mRNA splicing *in vitro. J. Biol. Chem.*, **260**, 14780.

88. Grabowski,P.J., Padgett,R.A. and Sharp,P.A. (1984) Messenger RNA splicing *in vitro*: an excised intervening sequence and a potential intermediate. *Cell*, **37**, 415.

89. Padgett,R.A., Konarska,M.M., Grabowski,P.J., Hardy,S.F. and Sharp,P.A. (1984) Lariat RNAs as intermediates and products in the splicing of messenger RNA precursors. *Science*, **225**, 898.

90. Krainer,A.R. and Maniatis,T. (1985) Multiple factors including the small nuclear ribonucleoproteins U1 and U2 are necessary for pre-mRNA splicing *in vitro. Cell*, **42**, 725.

91. Perkins,K.K., Furneaux,H.M. and Hurwitz,J. (1986) RNA splicing products formed with isolated fraction from HeLa cells are associated with fast-sedimenting complexes. *Proc. Natl. Acad. Sci. USA*, **83**, 887.

92. Cheng,S.-C. and Abelson,J. (1986) Fractionation and characterization of a yeast mRNA splicing extract. *Proc. Natl. Acad. Sci. USA*, **83**, 2387.

93. Lin,R.-J., Lustig,A.J. and Abelson,J. (1987) Splicing of yeast nuclear pre-mRNA *in vitro* requires a functional 40S spliceosome and several extrinsic factors. *Genes Dev.*, **1**, 7.

94. Reed,R. and Maniatis,T. (1985) Intron sequences involved in lariat formation during pre-mRNA splicing. *Cell*, **41**, 95.

95. Ruskin,B. and Green,M.R. (1985) Role of the 3′ splice site consensus sequence in mammalian pre-mRNA splicing. *Nature*, **317**, 732.

96. Newman,A.J., Lin,R.-J., Cheng,S.-C. and Abelson,J. (1985) Molecular consequences of specific intron mutations on yeast mRNA splicing *in vivo* and *in vitro. Cell*, **42**, 335.

97. Vijayraghavan,U., Parker,R., Tamm,J., Iimura,Y., Rossi,J., Abelson,J. and Guthrie,C. (1986) Mutations in conserved intron sequences affect multiple steps in the yeast splicing pathway, particularly assembly of the spliceosome. *EMBO J.*, **7**, 1683.

98. Fouser,L.A. and Friesen,J.D. (1986) Mutations in a yeast intron demonstrate the importance of specific conserved nucleotides for the two stages of nuclear mRNA splicing. *Cell*, **45**, 81.

99. Aebi,M., Hornig,H., Padgett,R.A., Reiser,J. and Weissmann,C. (1986) Sequence requirements for splicing of higher eukaryotic nuclear pre-mRNA. *Cell*, **47**, 555.

100. Hornig,H., Aebi,M. and Weissmann,C. (1986) Effect of mutations at the lariat branch acceptor site on β-globin pre-mRNA splicing *in vitro. Nature*, **324**, 589.

101. Freyer,G.A., Arenas,J., Perkins,K.K., Furneaux,H.M., Pick,L., Young,B., Roberts,R.J. and Hurwitz,J. (1987) *In vitro* formation of a lariat structure containing a $G^{2'}-^{5'}G$ linkage. *J. Biol. Chem.*, **262**, 4267.

102. Furneaux,H.M., Perkins,K.K., Freyer,G.A., Arenas,J. and Hurwitz,J. (1985) Isolation and characterization of two fractions from HeLa cells required for mRNA splicing *in vitro. Proc. Natl. Acad. Sci. USA*, **82**, 4351.

103. Krämer,A. and Keller,W. (1985) Purification of a protein required for the splicing of pre-mRNA and its separation from the lariat debranching enzyme. *EMBO J.*, **4**, 3571.

104. Reddy,R. and Busch,H. (1983) Small nuclear RNAs and RNA processing. *Prog. Nucleic Acid Res. Mol. Biol.,* **30,** 127.

105. Zhuang,Y. and Weiner,A.M. (1986) A compensatory base change in U1 snRNA suppresses a 5' splice site mutation. *Cell,* **46,** 827.

106. Black,D.L., Chabot,B. and Steitz,J.A. (1985) U2 as well as U1 small nuclear ribonucleoproteins are involved in pre-mRNA splicing *in vitro. Cell,* **42,** 737.

107. Berget,S.M. and Robberson,B.L. (1986) U1, U2, and U4/U6 small nuclear ribonucleoproteins are required for *in vitro* splicing but not polyadenylation. *Cell,* **46,** 691.

108. Black,D.L. and Steitz,J.A. (1986) Pre-mRNA splicing *in vitro* requires intact U4/U6 small nuclear ribonucleoprotein. *Cell,* **46,** 697.

109. Chabot,B., Black,D.L., LeMaster,D.M. and Steitz,J.A. (1985) The 3' splice site of pre-messenger RNA is recognized by a small nuclear ribonucleoprotein. *Science,* **230,** 1344.

110. Bindereif,A. and Green,M.R. (1986) Ribonucleoprotein complex formation during pre-mRNA splicing *in vitro. Mol. Cell. Biol.,* **6,** 2582.

111. Grabowski,P.J. and Sharp,P.A. (1986) Affinity chromatography of splicing complexes: U2, U5, and U4 + U6 small nuclear ribonucleoprotein particles in the spliceosome. *Science,* **233,** 1294.

112. Konarska,M.M. and Sharp,P.A. (1986) Electrophoretic separation of complexes involved in the splicing of precursors to mRNAs. *Cell,* **46,** 845.

113. Chabot,B. and Steitz,J.A. (1987) Multiple interactions between the splicing substrate and small nuclear ribonucleoproteins in spliceosomes. *Mol. Cell. Biol.,* **7,** 281.

114. Lustig,A.J., Lin,R.-J. and Abelson,J. (1986) The yeast RNA gene products are essential for mRNA splicing *in vitro. Cell,* **47,** 953.

115. Pikielny,C.W. and Rosbash,M. (1986) Specific small nuclear RNAs are associated with yeast spliceosomes. *Cell,* **45,** 869.

116. Wise,J.A., Tollervey,D., Maloney,D., Swerdlow,H., Dunn,E.J. and Guthrie,C. (1983) Yeast contains small nuclear RNAs encoded by single copy genes. *Cell,* **35,** 743.

117. Ares,M. (1986) U2 RNA from yeast is unexpectedly large and contains homology to vertebrate U4, U5, and U6 small nuclear RNAs. *Cell,* **47,** 49.

118. Riedel,N., Wise,J.A., Swerdlow,H., Mak,A. and Guthrie,C. (1986) Small nuclear RNAs from *Saccharomyces cerevisiae*: unexpected diversity in abundance, size, and molecular complexity. *Proc. Natl. Acad. Sci. USA,* **83,** 8097.

119. Parker,R., Siliciano,P.G. and Guthrie,C. (1987) Recognition of the TACTAAC box during mRNA splicing in yeast involves base pairing to the U2-like snRNA. *Cell,* **49,** 229.

120. Bass,B.L. and Cech,T.R. (1984) Specific interaction between the self-splicing RNA of *Tetrahymena* and its guanosine substrate: implications for biological catalysis by RNA. *Nature,* **308,** 820.

121. Jacquier,A. and Rosbash,M. (1986) Efficient trans-splicing of a yeast mitochondrial RNA group II intron implicates a strong 5' exon – intron interaction. *Science,* **234,** 1099.

122. Van der Veen,R., Arnberg,A.C. and Grivell,L.A. (1987) Self-splicing of a group II intron in yeast mitochondria: dependence on 5' exon sequences. *EMBO J.,* **6,** 1079.

123. Lerner,M.R., Boyle,J.A., Mount,S.M., Wolin,S.L. and Steitz,J.A. (1980) Are snRNPs involved in splicing? *Nature,* **283,** 220.

124. Rogers,J. and Wall,R. (1980) A mechanism for RNA splicing. *Proc. Natl. Acad. Sci. USA,* **77,** 1877.

125. Church,G.M., Slonimski,P.P. and Gilbert,W. (1979) Pleiotropic mutations within two yeast mitochondrial cytochrome genes block mRNA processing. *Cell,* **18,** 1209.

126. Anziano,P.Q., Hanson,D.K., Mahler,H.R. and Perlman,P.S. (1982) Functional domains in introns: *trans*-acting and *cis*-acting regions of intron 4 of the cob gene. *Cell,* **30,** 925.

127. De La Salle,H., Jacq,C. and Slonimski,P.P. (1982) Critical sequences within mitochondrial introns: pleiotropic mRNA maturases and *cis*-dominant signals of the box intron controlling reductase and oxidase. *Cell,* **28,** 721.

128. Weiss-Brummer,B., Rodel,G., Schweyen,R.J. and Kaudewitz,F. (1982) Expression of the split gene cob in yeast: evidence for a precursor of a 'maturase' protein translated from intron 4 and preceding exons. *Cell,* **29,** 527.

129. Nevins,J.R. (1983) The pathway of eukaryotic mRNA formation. *Annu. Rev. Biochem.,* **52,** 441.

130. Banerjee,A.K. (1980) 5' terminal cap structure in eukaryotic messenger ribonucleic acids. *Microbiol. Rev.*, **44**, 175.
131. Rhoads,R.E. (1985) The cap structure of eukaryotic messenger RNA and its interaction with cap-binding protein. *Prog. Mol. Subcell. Biol.*, **9**, 104.
132. Bunick,D., Zandomeni,R., Ackerman,S. and Weinmann,R. (1982) Mechanisms of RNA polymerase II-specific initiation of transcription *in vitro*: ATP requirement and uncapped runoff transcripts. *Cell*, **29**, 877.
133. Coppola,J.A., Field,A.S. and Luse,D.S. (1983) Promoter-proximal pausing by RNA polymerase II *in vitro*: transcripts shorter than 20 nucleotides are not capped. *Proc. Natl. Acad. Sci. USA*, **80**, 1251.
134. Furuichi,Y., Lafiandra,A. and Shatkin,A.J. (1977) 5'-Terminal structure and mRNA stability. *Nature*, **266**, 235.
135. Green,M.R., Maniatis,T. and Melton,D.A. (1983) Human β-globin pre-mRNA synthesized *in vitro* is accurately spliced in *Xenopus* oocyte nuclei. *Cell*, **32**, 681.
136. Noble,J.C.S., Prives,C. and Manley,J.L. (1986) *In vitro* splicing of simian virus 40 early pre-mRNA. *Nucleic Acids Res.*, **14**, 1219.
137. Ruskin,B. and Green,M.R. (1985) Specific and stable intron-factor interactions are established early during *in vitro* pre-mRNA splicing. *Cell*, **43**, 131.
138. Stevens,A and Maupin,M.K. (1987) A 5'-3' exoribonuclease of human placental nuclei: purification and substrate specificity. *Nucleic Acids Res.*, **15**, 695.
139. Patzelt,E., Blass,D. and Kuechler,E. (1983) CAP binding proteins associated with the nucleus. *Nucleic Acids Res.*, **11**, 5821.
140. Krainer,A.R. (1986) Nuclear pre-mRNA splicing *in vitro*. PhD Thesis, Harvard University, Cambridge, MA.
141. Konarska,M.M., Padgett,R.A. and Sharp,P.A. (1984) Recognition of cap structure in splicing *in vitro* of mRNA precursors. *Cell*, **38**, 731.
142. Edery,I. and Sonenberg,N. (1985) Cap-dependent RNA splicing in a HeLa nuclear extract. *Proc. Natl. Acad. Sci. USA*, **82**, 7590.
143. Patzelt,E., Thalmann,E., Hartmuth,K., Blaas,D. and Kuechler,E. (1987) Assembly of pre-mRNA splicing complex is cap-dependent. *Nucleic Acids Res.*, **15**, 1387.
144. Birnstiel,M.L., Busslinger,M. and Strub,K. (1985) Transcription termination and 3' processing: the end is in site! *Cell*, **41**, 349.
145. Moore,C.L. and Sharp,P.A. (1985) Accurate cleavage and polyadenylation of exogenous RNA substrate. *Cell*, **41**, 845.
146. Hart,P.R., McDevitt,M.A. and Nevins,J.R. (1985) Poly(A) site cleavage in a HeLa nuclear extract is dependent on downstream sequences. *Cell*, **43**, 677.
147. Zeevi,M., Nevins,J.R. and Darnell,J.E. (1981) Nuclear RNA is spliced in the absence of poly(A) addition. *Cell*, **26**, 39.
148. Keohavong,P., Gattoni,R., LeMoullec,J.M., Jacob,M. and Stevenin,J. (1982) The orderly splicing of the first three leaders of the adenovirus-2 major late transcript. *Nucleic Acids Res.*, **10**, 1215.
149. Nevins,J.R. and Darnell,J.E. (1978) Steps in the processing of Ad2 mRNA: poly(A)[+] nuclear sequences are conserved and poly(A) addition precedes splicing. *Cell*, **15**, 1477.
150. Wickens,M. and Stephenson,P. (1984) Role of the conserved AAUAAA sequence: four AAUAAA point mutations prevent messenger RNA 3' end formation. *Science*, **226**, 1045.
151. Moore,C.L. and Sharp,P.A. (1984) Site-specific polyadenylation in a cell-free reaction. *Cell*, **36**, 581.
152. Berget,S.M. (1984) Are U4 small nuclear ribonucleoproteins involved in polyadenylation? *Nature*, **309**, 179.
153. Hashimoto,C. and Steitz,J.A. (1986) A small nuclear ribonucleoprotein associates with the AAUAAA polyadenylation signal *in vitro*. *Cell*, **45**, 581.
154. Ryner,L.C. and Manley,J.L. (1987) Requirements for accurate and efficient mRNA 3' end cleavage and polyadenylation of a simian virus 40 early pre-RNA *in vitro*. *Mol. Cell. Biol.*, **7**, 495.
155. Pederson,T. (1983) Nuclear RNA – protein interactions and messenger RNA processing. *J. Cell Biol.*, **97**, 1321.
156. Choi,Y.D., Grabowski,P.J., Sharp,P.A. and Dreyfuss,G. (1986) Heterogeneous nuclear ribonucleoproteins: role in RNA splicing. *Science*, **231**, 1534.
157. Sierakowska,H., Szer,W., Furdon,P.J. and Kole,R. (1986) Antibodies to hnRNP core

proteins inhibit *in vitro* splicing of human β-globin pre-mRNA. *Nucleic Acids Res.,* **14**, 5241.

158. Nevins,J.R. (1979) Processing of the late adenovirus nuclear RNA to mRNA. Kinetics of formation of intermediates and demonstration that all events are nuclear. *J. Mol. Biol.,* **130**, 493.

159. Keohavong,P., Gattoni,R., Schmitt,P. and Stevenin,J. (1986) The different intron 2 species excised *in vivo* from the E2A pre-mRNA of adenovirus-2: an approach to analyze alternative splicing. *Nucleic Acids Res.,* **14**, 5207.

160. Sittler,A., Gallinaro,H. and Jacob,M. (1986) *In vivo* splicing of the pre-mRNAs from early region 3 of adenovirus-2: the products of cleavage at the 5' splice site of the common intron. *Nucleic Acids Res.,* **14**, 1187.

161. Roop,D.R., Nordstrom,J.L., Tsai,S.Y., Tsai,M.-J. and O'Malley,B.W. (1978) Transcription of structural and intervening sequences in the ovalbumin gene and identification of potential ovalbumin mRNA precursors. *Cell,* **15**, 671.

162. Zeitlin,S., Parent,A., Silverstein,S. and Efstratiadis,A. (1987) Pre-mRNA splicing and the nuclear matrix. *Mol. Cell. Biol.,* **7**, 111.

163. Pikielny,C.W. and Rosbash,M. (1985) mRNA splicing efficiency in yeast and the contribution of nonconserved sequences. *Cell,* **41**, 119.

164. Gruss,P., Lai,C.-J., Dhar,R. and Khoury,G. (1979) Splicing as a requirement for biogenesis of functional 16S mRNA of simian virus 40. *Proc. Natl. Acad. Sci. USA,* **76**, 4317.

165. Gruss,P. and Khoury,G. (1980) Rescue of a splicing defective mutant by insertion of an heterologous intron. *Nature,* **286**, 634.

166. Hamer,D.H. and Leder,P. (1979) Splicing and the formation of stable RNA. *Cell,* **18**, 1299.

167. Treisman,R., Novak,U., Favaloro,J. and Kamen,R. (1981) Transformation of rat cells by an altered polyoma virus genome expressing only the middle-T protein. *Nature,* **292**, 595.

168. Gruss,P., Efstratiadis,A., Karathanasis,S., Konig,M. and Khoury,G. (1981) Synthesis of stable unspliced mRNA from an intronless simian virus 40-rat preproinsulin gene recombinant. *Proc. Natl. Acad. Sci. USA,* **78**, 6091.

169. Yamada,Y., Avvedimento,V.E., Mudryj,M., Ohkubo,H., Vogeli,G., Irani,M., Pastan,I. and de Crombrugghe,B. (1980) The collagen gene: evidence for its evolutionary assembly by amplification of a DNA segment containing an exon of 54 bp. *Cell,* **22**, 887.

170. Wozney,J., Hanahan,D., Tate,V., Boedtker,H. and Doty,P. (1981) Structure of the proα2(I) collagen gene. *Nature,* **294**, 129.

171. Wieringa,B., Hoffer,E. and Weissmann,C. (1984) A minimal intron length but no specific internal sequence is required for splicing the large rabbit β-globin intron. *Cell,* **37**, 915.

172. Rautmann,G., Matthes,H.W.D., Gait,M.J. and Breathnach,R. (1984) Synthetic donor and acceptor splice sites function in an RNA polymerase B (II) transcription unit. *EMBO J.,* **3**, 2021.

173. Rautmann,G. and Breathnach,R. (1985) A role for branchpoints in splicing *in vivo*. *Nature,* **315**, 430.

174. Ruskin,B., Greene,J.M. and Green,M.R. (1985) Cryptic branchpoint activation allows accurate *in vitro* splicing of human β-globin intron mutants. *Cell,* **41**, 833.

175. Padgett,R.A., Konarska,M.M., Aebi,M., Hornig,H., Weissmann,C. and Sharp,P.A. (1985) Non-consensus branch-site sequences in the *in vitro* splicing of transcripts of mutant rabbit β-globin genes. *Proc. Natl. Acad. Sci. USA,* **82**, 8349.

176. Langford,C.J. and Gallwitz,D. (1983) Evidence for an intron-contained sequence required for the splicing of yeast RNA polymerase II transcripts. *Cell,* **33**, 519.

177. Pikielny,C.W., Teem,J.L. and Rosbash,M. (1983) Evidence for the biochemical role of an internal sequence in yeast nuclear mRNA introns: implications for U1 RNA and metazoan mRNA splicing. *Cell,* **34**, 395.

178. Domdey,H., Apostol,B., Lin,R.J., Newman,A., Brody,E. and Abelson,J. (1984) Lariat structures are *in vivo* intermediates in yeast pre-mRNA splicing. *Cell,* **39**, 611.

179. Rodriguez,J.R., Pikielny,C.W. and Rosbash,M. (1984) *In vivo* characterization of yeast mRNA processing intermediates. *Cell,* **39**, 603.

180. Cellini,A., Parker,R., McMahon,J., Guthrie,C. and Rossi,J. (1986) Activation of cryptic

TACTAAC box in the *Saccharomyces cerevisiae* actin intron. *Mol. Cell. Biol.,* **6**, 1571.
181. Cellini,A., Felder,E. and Rossi,J.J. (1986) Yeast pre-messenger RNA splicing efficiency depends on critical spacing requirements between the branchpoint and 3' splice site. *EMBO J.,* **5**, 1023.
182. Jacquier,A., Rodriguez,J.R. and Rosbash,M. (1985) A quantitative analysis of the effects of 5' junction and TACTAAC box mutants and mutant combinations on yeast mRNA splicing. *Cell,* **43**, 423.
183. Jacquier,A. and Rosbash,M. (1986) RNA splicing and intron turnover are greatly diminished by a mutant yeast branchpoint. *Proc. Natl. Acad. Sci. USA,* **83**, 5835.
184. Parker,R. and Guthrie,C. (1985) A point mutation in the conserved hexanucleotide at a yeast 5' splice junction uncouples recognition, cleavage and ligation. *Cell,* **41**, 107.
185. Fouser,L.A. and Friesen,J.D. (1987) Effects on mRNA splicing of mutations in the 3' region of the *Saccharomyces cerevisiae* actin intron. *Mol. Cell. Biol.,* **7**, 225.
186. Rymond,B.C. and Rosbash,M. (1985) Cleavage of 5' splice site and lariat formation are independent of 3' splice site in yeast mRNA splicing. *Nature,* **317**, 735.
187. Somasekhar,M.B. and Mertz,J.E. (1985) Exon mutations that affect the choice of splice sites used in processing the SV40 late transcripts. *Nucleic Acids Res.,* **13**, 5591.
188. Greenspan,D.S. and Weissmann,S.M. (1985) Synthesis of predominantly unspliced cytoplasmic RNAs by chimeric herpes simplex virus type 1 thymidine kinase-human β-globin genes. *Mol. Cell. Biol.,* **5**, 1894.
189. Ulfendahl,P.J., Pettersson,U. and Akusjarvi,G. (1985) Splicing of the adenovirus-2 E1A 13S mRNA requires a minimal intron length and specific intron signals. *Nucleic Acids Res.,* **13**, 6299.
190. Reed,R. and Maniatis,T. (1986) A role for exon sequences and splice site proximity in splice site selection. *Cell,* **46**, 681.
191. Eperon,L.P., Estibeiro,J.P. and Eperon,I.C. (1986) The role of nucleotide sequences in splice site selection in eukaryotic pre-messenger RNA. *Nature,* **324**, 280.
192. Fu,X.-Y. and Manley,J.L. (1987) Factors influencing alternative splice site utilization *in vivo*. *Mol. Cell. Biol.,* **7**, 738.
193. Klinz,F.J. and Gallwitz,D. (1985) Size and position of intervening sequences are critical for the splicing efficiency of pre-mRNA in the yeast *Saccharomyces cerevisiae*. *Nucleic Acids Res.,* **13**, 3791.
194. Chu,G. and Sharp,P.A. (1981) A gene chimaera of SV40 and mouse β-globin is transcribed and properly spliced. *Nature,* **289**, 378.
195. Treisman,R., Proudfoot,N.J., Shander,M. and Maniatis,T. (1982) A single-base change at a splice site in a β⁰-thalassemic gene causes abnormal RNA splicing. *Cell,* **29**, 903.
196. Horowitz,M., Cepko,C.L. and Sharp,P.A. (1983) Expression of chimeric genes in the early region of SV40. *J. Mol. Appl. Genet.,* **2**, 147.
197. Sharp,P.A. (1981) Speculations on RNA splicing. *Cell,* **23**, 643.
198. Chabot,B. and Steitz,J.A. (1987) Recognition of mutant and cryptic 5' splice sites by the U1 small nuclear ribonucleoprotein *in vitro*. *Mol. Cell. Biol.,* **7**, 698.
199. Tazi,J., Alibert,C., Temsamani,J., Reveillaud,I., Cathala,G., Brunel,C. and Jeanteur,P. (1986) A protein that specifically recognizes the 3' splice site of mammalian pre-mRNA introns is associated with a small nuclear ribonucleoprotein. *Cell,* **47**, 755.
200. Gerke,V. and Steitz,J.A. (1986) A protein associated with small nuclear ribonucleoprotein particles recognizes the 3' splice site of premessenger RNA. *Cell,* **47**, 973.
201. Iida,Y. (1985) Splice site signals of mRNA precursors are revealed by computer search. Site-specific mutagenesis and thalassemia. *J. Biochem.,* **97**, 1173.
202. Keller,W. (1984) The RNA lariat: a new ring to the splicing of mRNA precursors. *Cell,* **39**, 423.
203. Konarska,M.M., Grabowski,P.J., Padgett,R.A. and Sharp,P.A. (1985) Characterization of the branch site in lariat RNAs produced by splicing of mRNA precursors. *Nature,* **313**, 552.
204. Munroe,S.H. and Duthrie,R.S. (1986) Splice site consensus sequences are preferentially accessible to nucleases in isolated adenovirus RNA. *Nucleic Acids Res.,* **14**, 8447.
205. Solnick,D. (1985) Alternative splicing caused by RNA secondary structure. *Cell,* **43**, 667.

206. Lang,K.M. and Spritz,R.A. (1983) RNA splice site selection: evidence for a 5' to 3' scanning model. *Science,* **220**, 1351.

207. Kuhne,T., Wieringa,B. and Weissmann,C. (1983) Evidence against a scanning model of RNA splicing. *EMBO J.,* **2**, 727.

208. Krämer,A., Keller,W., Appel,B. and Lührmann,R. (1984) The 5' terminus of the RNA moiety of U1 small nuclear ribonucleoprotein particles is required for the splicing of messenger RNA precursors. *Cell.,* **38**, 299.

209. Dignam,J.D., Lebovitz,R.M. and Roeder,R.G. (1983) Accurate transcription initiation by RNA polymerase II in a soluble extract from isolated mammalian nuclei. *Nucleic Acids Res.,* **11**, 1475.

210. Reddy,R. (1986) Compilation of small RNA sequences. *Nucleic Acids Res.,* **14** (Suppl.), r61.

211. Theissen,H., Etzerodt,M., Reuter,R., Schneider,C., Lottspeich,F., Argos,P., Lührmann,R. and Philipson,L. (1986) Cloning of the human cDNA for the U1 RNA-associated 70K protein. *EMBO J.,* **5**, 3209.

212. Habets,W.J., Sillekens,P.T.G., Hoet,M.H., Schalken,J.A., Roebroek,A.J.M., Leunissen,J.A.M., Van de Ven,W.J.M. and Van Venrooij,W.J. (1987) Analysis of a cDNA clone expressing a human autoimmune antigen: full length sequence of the U2 small nuclear RNA-associated B″ antigen. *Proc. Natl. Acad. Sci. USA,* **84**, 2421.

213. Reddy,R., Henning,D. and Busch,H. (1985) Primary and secondary structure of U8 small nuclear RNA. *J. Biol. Chem.,* **260**, 10930.

214. Hinterberger,M., Pettersson,I. and Steitz,J.A. (1983) Isolation of small nuclear ribonucleoproteins containing U1, U2, U4, U5 and U6 RNAs. *J. Biol. Chem.,* **258**, 2604.

215. Bringmann,P., Rinke,J., Appel,B., Reuter,R. and Lührmann,R. (1983) Purification of snRNPs U1, U2, U4, U5 and U6 with 2,2,7-trimethylguanosine-specific antibody and definition of their constituent proteins reacting with anti-Sm and anti-(U1)RNP antisera. *EMBO J.,* **2**, 1129.

216. Mimori,T., Hinterberger,M., Petterson,I. and Steitz,J.A. (1984) Autoantibodies to the U2 small nuclear ribonucleoprotein in a patient with scleroderma-polymyositis overlap syndrome. *J. Biol. Chem.,* **259**, 560.

217. Bringman,P. and Lührmann,R. (1986) Purification of the individual snRNPs U1, U2, U5 and U4/U6 from HeLa cells and characterization of their protein constituents. *EMBO J.,* **5**, 3509.

218. Weinberg,R.A. and Penman,S. (1968) Small molecular weight monodisperse nuclear RNA. *J. Mol. Biol.,* **38**, 289.

219. Kunkel,G.R., Maser,R.L., Calvet,J.P. and Pederson,T. (1986) U6 small nuclear RNA is transcribed by RNA polymerase III. *Proc. Natl. Acad. Sci. USA,* **83**, 8575.

220. Reddy,R., Henning,D., Das,G., Harless,M. and Wright,D. (1987) The capped U6 small nuclear RNA is transcribed by RNA polymerase III. *J. Biol. Chem.,* **262**, 75.

221. Bringmann,P., Appel,B., Rinke,J., Reuter,R., Theissen,H. and Lührmann,R. (1984) Evidence for the existence of snRNAs U4 and U6 in a single ribonucleoprotein complex and for their association by intermolecular base pairing. *EMBO J.,* **3**, 1357.

222. Hashimoto,C. and Steitz,J.A. (1984) U4 and U6 RNAs coexist in a single small nuclear ribonucleoprotein particle. *Nucleic Acids Res.,* **12**, 3283.

223. Mount,S.M. and Steitz,J.A. (1981) Sequence of U1 RNA from *Drosophila melanogaster*: implications for U1 secondary structure and possible involvement in splicing. *Nucleic Acids Res.,* **9**, 6351.

224. Quigley,G.J. and Rich,A. (1976) Structural domains of transfer RNA molecules. The ribose 2' hydroxyl which distinguishes RNA from DNA plays a key role in stabilizing tRNA structure. *Science,* **194**, 796.

225. Mount,S.M., Petterson,I., Hinterberger,M., Karmas,A. and Steitz,J.A. (1983) The U1 small nuclear RNA-protein complex selectively binds a 5' splice site *in vitro. Cell,* **33**, 509.

226. Rinke,J., Appel,B., Blocker,H., Frank,R. and Lührmann,R. (1984) The 5' terminal sequence of U1 RNA complementary to the consensus 5' splice site of hnRNA is single stranded in intact U1 snRNP particles. *Nucleic Acids Res.,* **12**, 4111.

227. Yang,V.W., Lerner,M.R., Steitz,J.A. and Flint,S.J. (1981) A small nuclear

ribonucleoprotein is required for splicing of adenoviral early RNA sequences. *Proc. Natl. Acad. Sci. USA,* **78**, 1378.

228. Padgett,R.A., Mount,S.M., Steitz,J.A. and Sharp,P.A. (1983) Splicing of messenger RNA precursors is inhibited by antisera to small nuclear ribonucleoproteins. *Cell,* **35**, 101.

229. Donnis-Keller,H. (1979) Site-specific cleavage of RNA. *Nucleic Acids Res.,* **7**, 179.

230. Ohshima,Y., Itoh,M., Okada,N. and Miyata,T. (1981) Novel models for RNA splicing that involve a small nuclear RNA. *Proc. Natl. Acad. Sci. USA,* **78**, 4471.

231. Kinlaw,C.S., Robberson,B.L. and Berget,S.M. (1983) Fractionation and characterization of human small nuclear ribonucleoproteins containing U1 and U2 RNAs. *J. Biol. Chem.,* **258**, 7181.

232. Lelay-Taha,M.-N., Reveillaud,I., Sri-Widada,J., Brunel,C. and Jeanteur,P. (1986) RNA-protein organization of U1, U5 and U4-U6 small nuclear ribonucleoproteins in HeLa cells. *J. Mol. Biol.,* **189**, 519.

233. Tatei,K., Takemura,K., Mayeda,A., Fujiwara,Y., Tanaka,H., Ishihama,A. and Ohshima,Y. (1984) U1 RNA-protein complex preferentially binds to both 5' and 3' splice junction sequences in RNA or single-stranded DNA. *Proc. Natl. Acad. Sci. USA,* **81**, 6281.

234. Mayeda,A., Tatei,K., Kitayama,H., Takemura,K. and Ohshima,Y. (1986) Three distinct activities possibly involved in mRNA splicing are found in a nuclear fraction lacking U1 and U2 RNA. *Nucleic Acids Res.,* **14**, 3045.

235. Lührmann,B., Appel,B., Bringman,P., Rinke,J., Reuter,R. and Rothe,S. (1982) Isolation and characterization of rabbit anti-m^{2,2,7}G antibodies. *Nucleic Acids Res.,* **10**, 7103.

236. Smith,J.H. and Eliceiri,G.L. (1983) Antibodies elicited by N^2,N^2-7-trimethylguanosine react with small nuclear RNAs and ribonucleoproteins. *J. Biol. Chem.,* **258**, 4636.

237. Tollervey,D. and Mattaj,I.W. (1987) Fungal small nuclear ribonucleoproteins share properties with plant and vertebrate U-snRNPs. *EMBO J.,* **6**, 469.

238. Riedel,N., Wolin,S. and Guthrie,C. (1987) A subset of yeast snRNA's contains functional binding sites for the highly conserved Sm antigen. *Science,* **235**, 328.

239. Osheim,Y.N., Miller,O.L. and Beyer,A.L. (1985) RNP particles at splice junction sequences on *Drosophila* chorion transcripts. *Cell,* **43**, 143.

240. Kaltwasser,G., Spitzer,S.G. and Goldenberg,C.J. (1986) Assembly in an *in vitro* splicing reaction of a mouse insulin messenger RNA precursor into a 60–40S ribonucleoprotein complex. *Nucleic Acids Res.,* **14**, 3687.

241. Mayrand,S.H., Pedersen,N. and Pederson,T. (1986) Identification of proteins that bind tightly to pre-mRNA during *in vitro* splicing. *Proc. Natl. Acad. Sci. USA,* **83**, 3718.

242. Pikielny,C.W., Rymond,B.C. and Rosbash,M. (1986) Electrophoresis of ribonucleoproteins reveals an ordered assembly pathway of yeast splicing complexes. *Nature,* **324**, 341.

243. Gilbert,W., Maxam,A. and Mirzabekov,A.D. (1976) Contacts between the LAC repressor and DNA revealed by methylation. In *Control of Ribosome Synthesis.* Kjelgaard,N.O. and Maaloe,O. (eds), Academic Press, New York, p. 139.

244. Galas,D. and Schmitz,A. (1978) DNase footprinting: a simple method for the detection of protein-DNA binding specificity. *Nucleic Acids Res.,* **5**, 3157.

245. Rymond,B.C. and Rosbash,M. (1986) Differential nuclease sensitivity identifies tight contacts between yeast pre-mRNA and spliceosomes. *EMBO J.,* **5**, 3517.

246. Laski,F.A., Rio,D.C. and Rubin,G.M. (1986) Tissue specificity of *Drosophila* P element transposition is regulated at the level of mRNA splicing. *Cell,* **44**, 7.

247. Szostak,J.W. (1986) Enzymatic activity of the conserved core of a group I self-splicing intron. *Nature,* **322**, 83.

248. Solnick,D. (1985) *Trans* splicing of mRNA precursors. *Cell,* **42**, 157.

249. Konarska,M.M., Padgett,R.A. and Sharp,P.A. (1985) *Trans* splicing of mRNA precursors *in vitro*. *Cell,* **42**, 165.

250. Borst,P. (1986) Discontinuous transcription and antigenic variation in trypanosomes. *Annu. Rev. Biochem.,* **55**, 701.

251. Van der Ploeg,L.H.T. (1986) Discontinuous transcription and splicing in trypanosomes. *Cell,* **47**, 479.

252. Laird,P.W., Zomerdijk,J.C.B.M., De Korte,D. and Borst,P. (1987) *In vivo* labeling

of intermediates in the discontinuous synthesis of mRNAs in *Trypanosoma brucei*. *EMBO J.*, **6**, 1055.

253. Murphy,W.J., Watkins,K.P. and Agabian,N. (1986) Identification of a novel Y branch structure as an intermediate in trypanosome mRNA processing: evidence for *trans* splicing. *Cell*, **47**, 517.

254. Sutton,R.E. and Boothroyd,J.C. (1986) Evidence for *trans* splicing in trypanosomes. *Cell*, **47**, 527.

255. Tschudi,C., Richards,F.F. and Ullu,E. (1986) The U2 RNA analogue of *Trypanosoma brucei* gambiense: implications for a splicing mechanism in trypanosomes. *Nucleic Acids Res.*, **14**, 8893.

256. Fromm,H., Edelman,M., Koller,B., Goloubinoff,P. and Galun,E. (1986) The enigma of the gene coding for ribosomal protein S12 in the chloroplasts of *Nicotiana*. *Nucleic Acids Res.*, **14**, 883.

257. Fukuzawa,H., Kohchi,T., Shirai,H., Ohyama,K., Umesono,K., Inokuchi,H. and Ozeki,H. (1986) Coding sequences for chloroplast ribosomal protein S12 from the liverwort, *Marchantia polymorpha*, are separated far apart on the different DNA strands. *FEBS Lett.*, **198**, 11.

258. Torazawa,K., Hayashida,N., Obokata,J., Shinozaki,K. and Sugiura,M. (1986) The 5' part of the gene for ribosomal protein S12 is located 30 kbp downstream from its 3' part in tobacco chloroplast genome. *Nucleic Acids Res.*, **14**, 3143.

259. Koller,B., Fromm,H., Galun,E. and Edelman,M. (1987) Evidence for *in vivo trans* splicing of pre-mRNAs in tobacco chloroplasts. *Cell*, **48**, 111.

260. Berget,S.M., Moore,C. and Sharp,P.A. (1977) Spliced segments at the 5' terminus of adenovirus 2 late mRNA. *Proc. Natl. Acad. Sci. USA*, **74**, 3171.

261. Chow,L.T., Gelinas,T.E., Broker,T.R. and Roberts,R.J. (1977) An amazing sequence arrangement at the 5' ends of adenovirus 2 messenger RNA. *Cell*, **12**, 1.

262. Kaine,B.P., Gupta,R. and Woese,C.R. (1983) Putative introns in tRNA genes of prokaryotes. *Proc. Natl. Acad. Sci. USA*, **80**, 3309.

263. Sharp,P.A. (1985) On the origin of RNA splicing and introns. *Cell*, **42**, 397.

264. Daniels,C.J., Gupta,R. and Doolittle,W.F. (1985) Transcription and excision of a large intron in the tRNA[trp] gene of an archaebacterium, *Halobacterium volcanii*. *J. Biol. Chem.*, **260**, 3132.

265. Gilbert,W. (1986) The RNA world. *Nature*, **319**, 618.

266. Kjems,J. and Garret,R.A. (1985) An intron in the 23S ribosomal RNA gene of the archaebacterium *Desulfurococcus mobilis*. *Nature*, **318**, 675.

267. Epstein,L.M. and Gall,J.G. (1987) Self-cleaving transcripts of satellite DNA from the newt. *Cell*, **48**, 535.

268. Walter,P. and Blobel,G. (1982) Signal recognition particle contains a 7S RNA essential for protein translocation across the endoplasmic reticulum. *Nature*, **299**, 691.

269. Korneeva,G.A., Petrova,A.N., Venkstern,T.V. and Bayev,A.A. (1979) Primary structure of the nucleic acid from the 1,4-α-glucan branching enzyme. *Eur. J. Biochem.*, **96**, 339.

270. Wong,T.W. and Clayton,D.A. (1986) DNA primase of human mitochondria is associated with structural RNA that is essential for enzymatic activity. *Cell*, **45**, 817.

271. Sakai,H. (1967) A ribonucleoprotein which catalyzes thiol-disulfide exchange in the sea urchin egg. *J. Biol. Chem.*, **242**, 1458.

272. Huang,D.-D., Wang,W.Y., Gough,S.P. and Kannangara,C.G. (1984) δ-Aminolevulinic acid-synthesizing enzymes need an RNA moiety for activity. *Science*, **225**, 1482.

273. Arnberg,A.C., Van der Horst,G. and Tabak,H.F. (1986) Formation of lariats and circles in self-splicing of the precursor to the large ribosomal RNA of yeast mitochondria. *Cell*, **44**, 235.

274. Jacquier,A. and Dujon,B. (1985) An intron-encoded protein is active in a gene conversion process that spreads an intron into a mitochondrial gene. *Cell*, **41**, 383.

275. Macreadie,I.G., Scott,R.M., Zinn,A.R. and Butow,R.A. (1985) Transposition of an intron in yeast mitochondria requires a protein encoded by that intron. *Cell*, **41**, 395.

276. Crick,F. (1979) Split genes and RNA splicing. *Science*, **204**, 264.

277. Cavalier-Smith,T. (1985) Selfish DNA and the origin of introns. *Nature*, **315**, 283.

278. Doolittle,W.F. (1978) Genes in pieces: were they ever together? *Nature*, **272**, 581.

279. Darnell,J.E. (1978) Implications of RNA-RNA splicing in evolution of eukaryotic cells. *Science*, **202**, 1257.

280. Gargouri,A., Lazowska,J. and Slonimski,P.P. (1983) DNA-splicing of introns in a gene: a general way of reverting intron mutations. In *Mitochondria 1983*. Schweyen,R.J., Wolf,K. and Kaudewitz,F. (eds), Walter de Gruyter, Berlin and New York, p. 259.
281. Gilbert,W. (1978) Why genes in pieces? *Nature*, **271**, 501.
282. Gilbert,W. (1985) Genes-in-pieces revisited. *Science*, **228**, 823.
283. Gilbert,W., Marchionni,M. and McKnight,G. (1986) On the antiquity of introns. *Cell*, **46**, 151.
284. Rogers,J. (1985) Exon shuffling and intron insertions in serine protease genes. *Nature*, **315**, 458.
285. Ng,R., Domdey,H., Larson,G., Rossi,J.J. and Abelson,J. (1985) A test for intron function in the yeast actin gene. *Nature*, **314**, 183.
286. Bozzoni,I., Fragapane,P., Annesi,F., Pierandrei-Amaldi,P., Amaldi,F. and Beccari,E. (1984) Expression of two *Xenopus laevis* ribosomal protein genes in injected frog oocytes. *J. Mol. Biol.*, **180**, 987.
287. Dabeva,M.D., Post-Beittenmiller,M.A. and Warner,J.R. (1986) Autogenous regulation of splicing of the transcript of a yeast ribosomal protein gene. *Proc. Natl. Acad. Sci. USA*, **83**, 5854.
288. Johnson,P.F. and Abelson,J. (1983) The yeast tRNA[Tyr] gene intron is essential for correct modification of its tRNA product. *Nature*, **302**, 681.
289. Banerji,J., Olson,L. and Schaffner,W. (1983) A lymphocyte-specific cellular enhancer is located downstream of the joining region in immunoglobulin heavy chain genes. *Cell*, **33**, 729.
290. Gillies,S.D., Morrison,S.L., Oi,V.T. and Tonegawa,S. (1983) A tissue-specific transcription enhancer element is located in the major intron of a rearranged immunoglobulin heavy chain gene. *Cell.*, **33**, 717.
291. Queen,C. and Stafford,J. (1984) Fine mapping of an immunoglobulin gene activator. *Mol. Cell. Biol.*, **4**, 1042.
292. Henikoff,S., Keene,M.J.A., Fechtel,K. and Fristrom,J.W. (1986) Gene within a gene: nested *Drosophila* genes encode unrelated proteins on opposite DNA strands. *Cell*, **44**, 33.
293. King,C.R. and Piatigorsky,J. (1983) Alternative RNA splicing of the murine αA-crystallin gene: protein-coding information within an intron. *Cell*, **32**, 707.
294. Dodgson,J.B. and Engel,J.D. (1983) The nucleotide sequence of the adult chicken α-globin genes. *J. Biol. Chem.*, **258**, 4623.
295. Erbil,C. and Niessing,J. (1983) The primary structure of the duck α^D-globin gene: an unusual 5' splice junction sequence. *EMBO J.*, **2**, 1339.
296. Citri,Y., Colot,H.V., Jacquier,A.C., Yu,Q., Hall,J.C., Baltimore,D. and Rosbash,M. (1987) A family of unusually spliced biologically active transcripts encoded by a *Drosophila* clock gene. *Nature*, **326**, 42.
297. Lowery,D.E. and Van Ness,B.G. (1987) *In vitro* splicing of kappa immunoglobulin precursor mRNA. *Mol. Cell. Biol.*, **7**, 1346.
298. Schatz,P.J., Pillns,N., Grisafi,P., Solomon,F. and Botstein,D. (1986) Two functional α-tubulin genes of the yeast *Saccharomyces cerevisiae* encode divergent proteins. *Mol. Cell. Biol.*, **6**, 3711.
299. Molenaar,C.M.T., Wondt,L.P., Jansen,A.E.M., Mager,W.H., Planta,R.J., Donovan, D.M. and Pearson,N.J. (1984) Structure and organization of two linked ribosomal protein genes in yeast. *Nucleic Acids Res.*, **12**, 7345.
300. Nieuwint,R.T.M., Molenaar,C.M.T., Van Bommel,J.H., Van Raamsdonk-Duin,M.M.C., Mager,W.H. and Planta,R.J. (1985) The gene for yeast ribosomal protein S31 contains an intron in the leader sequence. *Curr. Genet.*, **10**, 1.
301. Simon,M., Seraphin,B. and Faye,G. (1986) KIN28, a yeast split gene coding for a putative protein kinase homologous to CDC28. *EMBO J.*, **5**, 2697.

Index